OPEN MARXISM

OPEN MARXISM

Edited by
Werner Bonefeld, Richard Gunn
and Kosmas Psychopedis

VOLUME II

THEORY AND PRACTICE

PLUTO PRESS

First published 1992 by Pluto Press
345 Archway Road, London N6 5AA

A catalogue record for this book
is available from the British Library

Library of Congress Cataloging in Publication Data
Open Marxism/edited by Werner Bonefeld, Richard Gunn,
and Kosmas Psychopedis.
390 p. 22 cm.
Includes bibliographical references and indexes.
Contents: v. 1. Dialectics and history – v. 2. Theory and
practice.

ISBN 9780745305912 (v.1)   ISBN 9780745304250 hb (v.2):
ISBN 0745304249 (v.1)   ISBN 0745304257 hb (v.2):

I. Philosophy, Marxist. I. Bonefield, Werner, 1960–
II. Gunn, Richard, 1947– . III. Psychopedis, Kosmas.
BF809.8.058   1991
335.4'11 – dc20   91-8323
                                                       CIP

Typeset by Centracet, Cambridge

Printed and bound by Antony Rowe Ltd, Eastbourne

Contents

Acknowledgements

We should like to extend thanks to Michael Hardt who translated the article by Negri. Grants supplied by the University of Edinburgh made possible the coordination of our editorial work.

Notes on Contributors

Werner Bonefeld, lecturer in Politics at the University of Edinburgh, is the author of 'The Reformulation of State Theory', *Capital & Class*, no. 33 (1987); 'Open Marxism', *Common Sense*, no. 1 (1987); and is co-editor (with John Holloway) of *Post-Fordism and Social Form* (London, 1991). Werner Bonefeld is actively involved in the Conference of Socialist Economists, has been a member of the editorial board of *Capital & Class* and is a member of the editorial collection of *Common Sense*.

Harry Cleaver holds a professorship in the Department of Economics, University of Texas at Austin. A noted proponent of autonomist thought in the United States, he is the author of 'Food, Famine and the International Crisis', *Zerowork*, no. 2 (1977); 'Supply-side Economics: Splendori e miserie', *Metropoli* no. 7 (1981); 'Reaganisme et rapports de classe aux États-Unis', in M.–B. Tahon and A. Corten (eds), *L'Italie: le philosophe et le gendarme* (Montreal, 1986; 'The Uses of an Earthquake', *Common Sense*, no. 8, Edinburgh (1989); 'Close the IMF, Abolish Debt and End Development: A Class Analysis of the International Debt Crisis', *Capital & Class*, no. 39, London (1989). His *Reading Capital Politically* (Austin, Texas) appeared in 1979 and he has been involved in the translation of Antonio Negri's work into English. Harry Cleaver has been an editor of *Zerowork*.

Joseph Fracchia studied history and philosophy at the University of California, Davis. He also spent two years studying social theory at Georg August University in Göttingen. Since 1986, he has been teaching at the University of Oregon. His book *Die Marxsche Aufhebung der Philosophie, und der Philosophische Marxismus* appeared in 1987.

Richard Gunn lectures in political theory at the Department of Politics, University of Edinburgh. He is the author of 'Marxism and Ideas of Power and Participation', in J. Bloomfield (ed.), *Class, Hegemony and Party*, (London, 1977); 'Is Nature Dialectical?', *Marxism Today*, vol. 21, no. 2, London (1977); 'Practical Reflexivity in Marx' and 'Marxism and Mediation', *Common Sense*, nos 1 and 2, Edinburgh (1987); 'Marxism and Philosophy: A Critique of Critical Realism', *Capital & Class*, no. 37, London (1989); 'Marxism, Metatheory and Critique', in W. Bonefeld/J. Holloway (eds), *Post-Fordism and Social Form* (London 1991); 'Reclaiming Experience', *Science as Culture* no 11 (1991) London, A recent publication is 'Marxism and common sense', *Common Sense*, no. 11, Edinburgh 1991. Richard Gunn is actively involved in the Conference of Socialist Economists, has been a member of the editorial boards of *Capital & Class* and is a member of the editorial collective of *Common Sense*.

John Holloway lectures in the Department of Politics, University of Edinburgh. His publications include 'The Red Rose of Nissan', *Capital & Class*, no. 32, London (1987); 'Learning to Bow, Post-Fordism and Technological Determinism' (with E. Peláez), *Science as Culture*, London (1989); 'State as Class Practice', *Research in Political Economy*, vol. 3, (1980); *State and Capital: A Marxist Debate* (edited with S. Picciotto) (London, 1978); *In and Against the State* (with others, as the London Edinburgh Weekend Return Group) (London 1980); *Post-Fordism and Social Form* (edited with W. Bonefeld)(London, 1991). John Holloway has been actively involved in the Conference of Socialist Economists and has taught as a guest professor at the University of Mexico.

Antonio Negri held a professorship of philosophy of law and state doctrine at the University of Padova, where he founded the Institute of Political Science. During the 1960s and 1970s he was a member of the editorial boards of autonomist journals. In 1977, forced to leave Italy following an investigation into his thought and editorial activities, Negri went to Paris where he was named professor at the University of Paris VII and at the École Normale Superième. In 1979, during the course of a second investigation, Negri was accused of being the 'brains' of Italian terrorism. After four and a half years of detention without trial in special prisons, Negri was elected as deputy to the Italian parliament. According to Italian law, he was released from prison, but, several months later, the parliament revoked his immunity and he once again fled to France. In his absence, he was condemned to 30 years' imprisonment. The conditions of his trial have been criticised by Amnesty International. Presently, Negri teaches political science at the University of Paris VIII. English-language translations of Negri's work include: *Marx Beyond Marx* (South Hadley, 1984); *Revolution Retrieved* (London, 1988); *The Politics of Subversion* (Cambridge, 1989); *The Savage Anomaly: The Power of Spinoza's Metaphysics and Politics* (Minnesota, 1990).

Kosmas Psychopedis studied at the Universities of Athens and Frankfurt and has been a professor of Political Science at the University of Göttingen and at the Panteios School of Political Science in Athens. He currently holds a professorship in the Department of Economics at the University of Athens. His publications include 'Die Möglichkeit der Gesellschaftsphilosophie bei Hegel', *Gesellschaft: Beiträge zur Marxschen Theorie*, vol. 5 (Frankfurt a.M., 1975); 'Notes on Mediation-Analysis', *Common Sense*, no. 5 (1988); 'Crisis of Theory in the Contemporary Social Sciences', in W. Bonefeld and J. Holloway (eds), *Post-Fordism and Social Form* (London, 1991); *Untersuchungen zur politischen Theorie I. Kants* (Göttingen, 1980); and *Geschichte und Methode* (Frankfurt a.M/ New York, 1984).

Cheyney Ryan attended Harvard University, from which he was expelled for political activity, and Boston University, from which he was also expelled for such activity and then readmitted to receive his Ph.D. in 1974. Since then he has taught at the University of Oregon, where he is currently Professor of Philosophy and Head of the Peace Studies Programme. He has published in philosophy, economics, sociology and law journals. His recent publications include 'Liberalism – Equality and Exploitation', *Revue Internationale de Philosophy*, Paris (1988); and 'Recognition and Visions of Equality', *Georgia Law Review* (1990). He is also a playwright whose work has been performed throughout the United States.

Introduction

WERNER BONEFELD, RICHARD GUNN, KOSMAS PSYCHOPEDIS

The present volume continues where our first left off: we turn from the notion of dialectics and history to the notion of the unity of theory and pratice. Theory and practice cannot be separated from the open Marxist debate on history and dialectics; both presuppose and are the result of each other. Such was implied in our introduction to Volume One where we outlined the notion of an open Marxism. In the present introduction we take the 'definition' of open Marxism for granted and launch directly into addressing the issue of theory and practice within the open Marxist debate. The aim of the volume is to elucidate the relationship between theory and practice and to explore some of the issues to which it gives rise. These issues include the epistemological foundations of Marxist theory (Gunn/Fracchia and Ryan), class and self-determination (Cleaver/Negri) and fetishism and class composition (Holloway). None of our contributors would be likely to agree with each other on the precise understanding of the unity of theory and practice, nor for that matter on the relation between structure and struggle upon which the notion of the unity of theory and practice turns. However, a common concern of our contributors is their rejection of an understanding of practice as merely attendant upon the unfolding of structural or deterministic 'laws'. This common concern might be summed up in terms of an understanding of class as the constitutive power of history and of commitment as a requirement for taking social responsibility.

In our introduction to the first volume we emphasised that open Marxism entails the openness of categories themselves. The openness of categories – an openness on to practice – obtains as a reflexive critique of ideologies and social phenomena, which, for their part, exist as moments of historically asserted forms of class struggle. Open Marxism's starting point is the class antagonism between capital and labour. An understanding of the 'primacy of class' implies a constant change on the part of social 'reality' and a constant change in the form of the class struggle. In turn, the understanding of social reality as constantly moving implies the incompleteness of categories as the social development appears in various forms and within changing empirical circumstances. Instead of the theoretical certainty of a Marxism of dogmatic closure, open Marxism reclaims the incompleteness of the process of thinking and readopts the unpredictability of the 'legitimation of chance'[1] i.e. the unpredictability of the movement of class struggle. Following upon the contributions in Volume One, the understanding of social objectivity as alienated subjectivity entails an internal relation, rather than an external dualism, between structure and struggle.

The separation between structure and struggle entails a deterministic conceptualisation of capital in that capital becomes a structure of inescapable lines of development, subordinating social practice to predetermined 'laws'. On the other hand, understanding capital as a social relation implies that there are no inescapable lines of development. Alleged 'lines of development' are the fetished forms of the capital-labour relation itself, i.e. of class struggle. Open Marxism insists on the antagonistic nature of social existence. This being so, the Marxist understanding of a unity of theory and practice entails *not* the theoretical suppression of class struggle, but the invocation of class struggle as the movement of the contradiction in which capital, itself, consists.

All of this carries with it pungent implications for the way in which Marxism approaches the issue of 'class'. If there were to be a dualism between structure and struggle, then nothing less than a one-sided abstraction of capital would obtain. Such abstraction, in turn, assumes that the logic of capital is the key to the emergence and development of the working class. Determinist Marxism and capital-logic Marxism have marched together, whether in Leninist-revolutionary or in reformist or in post-fordist guise. Capital sets the questions, and it is up to the working class to propose whatever answers it sees fit.

Within this tradition, 'class' is construed in sociological terms. Debates on the score of whether the proletariat is to be identified with the manual working class, on the score of whether the manual working class still counts as the majority of workers, and on attempts to identify a 'new working class' or a 'new petty bourgeoisie'[2] all take as their basic problematic the question: *to which class* can this or that individual be assigned? More recent versions of the same problematic have foregrounded the notion of 'contradictory' class relations,[3] but this renders the problematic more complex without challenging its basic features. The notion of classes as pigeonholes or 'locations' to which the sociologist must assign individuals ultimately invokes static and struggle-disconnected structures. Current fordist/post-fordist debates about the class significance of work at computer terminals or in the service industries are to the same effect. These are quantitative conceptions of class, presupposing that the political significance of class can be established by counting heads. The alternative qualitative conception of class, which addresses it not as matter of grouping individuals but as a contradictory and antagonistic social relation, has hitherto been a somewhat marginalised tradition of Marxist thought.

This latter tradition has always insisted that social phenomena have to be seen as *forms* assumed by class struggle, as forms in and against which social conflict obtains. Capital, it suggests, is a social relation of an antagonistic kind. Capital is therefore always in a position of having to recompose itself by reintegrating the working class into the capital-relation. The conceptual foundation for this approach is the circumstance that the relation between capital and labour is asymmetrical: capital depends upon labour, for its valorisation, but labour for its part in no way depends, necessarily, on capital's rule. The political foundation for such an approach is to be discovered in the history of class struggle – frequently ignored or marginalised, even by Marxist theoreticians – over the extraction of surplus labour which has occurred ever since capital*ism*, as a reified social 'system', got into gear.

The question of 'form', understood as 'mode of existence', was discussed in Volume One: the question addressed in the present volume is that of the implications for class and the unity of theory and practice of this understanding of 'form'.

Open Marxism urges both the opening of concepts on to practice, whose capacity for renewal and innovation always surprises us, and the mediating of that practice through categories of a critical and self-critical kind. Thereby it transcends the dichotomy: theory *or* practice. The notion that theory and practice form a unity is as old as Marxism itself; however, traditional schools of Marxism – and these include the dialectical materialist Marxism-Leninism of the Stalin years as well as the 'structuralism' of more recent decades – have tended to see the theorist as standing outside of society and as reflecting, externally, upon it. Within such conceptual frameworks, the unity of theory and practice can amount only to the *application* of theory to practice. Structuralism and voluntarism are dichotomous, though conjoint, outcomes of such an approach. Structuralism and voluntarism are complementary inasmuch as they are the result of the separation between allegedly abstract laws and subjectivity. Open Marxism moves beyond such a dichotomy by acknowledging theory to obtain in and of practice and by acknowledging practice (that is, human or social practice) to occur only in some reflectively considered, or unreflectively assumed, set of terms. Theory can be no less concrete than practice, and practice can be no less abstract than theory. We do not have two movements dualistically counter-posed but a single theoretico-practical class movement which, to be sure, contains differences and diversity within itself.

A single theoretico-practical class movement of this kind entails the practical reflexivity of theory and the theoretical reflexivity of practice as different moments of the same totalisation. If there were to be, as structuralist approaches urge, a disunity of subjectivity and objectivity, then there could be no question of an internal relation between social phenomena or of, what is the same thing, social moments subsisting as one another's mode of existence (or 'form'). Hence, form-analysis and the unity of theory and practice imply one another. The internal relation between theory and practice, or between subjectivity and objectivity, connotes the theoretical diffi-culties, and failures, of structuralist approaches: first of all capital is seen by such approaches as a logical construct, moving within a particular set of objective laws, whereas, secondly, any historical analysis renders necessary the reintroduction of subjective aspects. Hence, the apparently opposed terms 'structuralism' and 'voluntarism' stand related to one another in a complementary way.

The interlinked themes of form-analysis and the unity of theory and practice introduce us, directly, on to the terrain of critique. Precisely the dualistic separations of a fetished world – the separa-tions of subject from object, of struggle from structures, of theory from practice and of one 'region' of society from another – are to be called in question rather than being inscribed, in taken-for-granted fashion, as the principles of social thought. Conversely, critique implies form-analysis and the thesis of the unity of theory and practice. It does so because only if social forms (including theory's own form) are understood as internally related modes of existence can the fetishism of discrete 'regions' and 'entities' and 'facts' and 'ideologies' be called to account.

This approach to critique allows us to see what is wrong with various allegedly critical schools of contemporary thought. The case of structuralism has already been discussed: it premises itself upon a distinction between regions of social existence (e.g. 'the economic' and 'the political'), thereby inscribing as a methodological principle the fetishism which criticism contends against. A related case is that of 'Critical Realism', which (cf. Psychopedis in Volume One) employs Kantian transcendental argument in such a way as to consolidate rather than criticise the phenomena it explores.[4] The 'Rational Choice' or 'Analytical' Marxism of the 1980s, similarly, takes for granted a conception of bourgeois individuality *and* a politics/economics separation *à la* structuralism.[5] Two further schools of thought which characterise themselves as critical share similar deficiencies. Post-modernism declares against 'teleological' conceptions of history but places itself at the mercy of, precisely, history by declaring virtually all historical analysis to be of a teleological sort.[6] It draws its own historical teeth. Philosophy of science, which, like post-modernism, has always stood opposed to historical teleologies,[7] has in recent years abutted on to history[8] but remains equivocal as between philosophical and merely historical accounts.

What is lacking in all of these approaches is that they fail to deepen critique into a theory-practice unity. They advertise theorising which is either merely theory *of* practice (as in structuralism) or merely theory which is *in* practice (e.g. post-modernism). At most, as in the Critical Realism version of philosophy of science, they see theory and practice as causally (and, therefore, still externally) linked. What lies beyond their horizons is a conception of theory as *in and of* practice, i.e. which is analytical and critical and social-scientific and philosophical all in the same movement and in the same breath. Only a *practically* reflexive critique can achieve this. Or, in other words, critique can be achieved only in the light of the category of form. *Form*, once freed from the grip of structuralism and empiricism which understands it merely in terms of the different species something or other can take, signals the mode(s) of existence of the contradictory movement in which social existence consists. Critique moves within its object and, at the same time, is a moment of its object. Thus, critique implies a unity of theory and practice which permits a demystification of 'structures', 'empirical facts' and 'ideologies' as fetished forms assumed by social relations. The forms assumed by social relations are the object of critique which, itself, is an open process: there is no externality to the form(s) assumed by social relations. On the same score, critique is essentially practical as it theorises the form-giving fire of social relations. Critique implies form-analysis and *vice-versa*. 'All social relations are essentially practical' (Marx).

Richard Gunn, whose recent strongly debated critique of Critical Realism maps out a fresh approach to questions of Marxist method-ology, seeks to renew the tradition of a theory/practice-based Marx-ism stemming from, amongst others, Lukács and to deploy the insights of such a tradition against historical materialist thought in both its old-style and its new-style versions. Gunn's conception of the unity of theory and higher-order metatheory criticises versions of 'Marxism' which distinguish between structure and subjectivity. His contention is that once we attempt a theory *of* history and society we sever, at source, the unity of theory and practice which Marx projects.

Joseph Fracchia, whose recent work on Marxism and philosophy reconstructs the conception of critique, and *Cheyney Ryan*, who has published on the themes of equality and exploitation, pick up on the question of the unity of theory and practice by developing an argument the context for which is twofold: post-modern thought and contemporary developments in philosophy of science. Their intriguing contention is that the work of Thomas Kuhn can help us to overcome not only the deficiencies of post-modernism, but also the separation of theory from practice inherent in philosophy-of-science style of thought. Their reformulation of Kuhn allows them to bring the practical category of 'commitment' on to centre-stage. Theoretical commitment and practical openness go hand in hand.

Antonio Negri, who is known for his work on Marx and class and who has been a leading theorist within the autonomist Marxist tradition, rejects a sociology of class cast in terms of capitalist reproduction. Negri's stress is on the power of labour as constitutive of social activity; this emphasis allows him to construe social development in terms of revolutionary constitution and rupture. In Negri's view, communism as a constitutive power obtaining (already) within, and against, capital is the requisite category for any theory purporting to focus on issues of class. The possibility of a constitution of communism he discusses in terms of the category of 'value'. Contrary to seeing value as an economistic 'measure', Negri stresses value's qualitative and political dimension. Negri's view of value is at the opposite pole from that favoured by Marxist 'economics' and 'political science'. Contrary to these views, Negri stresses the real and radical possibility of communist constituent power opened up at the present time.

Harry Cleaver, who has published extensively on the political dimension of Marx's *Capital* and on the constitutive power of labour, works within the autonomist marxist tradition associated with Panzieri, Tronti and Negri. In his contribution he contends that 'new social movements' need to be understood in terms of class. He focuses on the self-determinating struggle against capitalist appropriation and destruction of social life. The categories of Marxism are discussed in terms of 'inversion', or double-sidedness. Crucially, the inverse side of valorisation is 'disvalorisation', i.e. loss of identity and the destruction of traditional values (in the plural) as a consequence of capital's parasitic appropriation of creativity. Resistance to this appropriation involves 'self-valorisation' – that is, the emancipatory project of the working class.

John Holloway, who has published widely on the state and class practice, offers the category of 'fetishism' as allowing us to understand the internal relation between 'value' and class struggle. Social structures become fetished in the same movement as they are studied separately from the struggle in and through which they subsist. Theory becomes fetished in the same movement as it fails to recognise its inherence in a practical world, and practice as understood by such theory is construed in a fetished – a 'structuralist' – way. The critique of fetishism and the thesis of a unity of theory and practice, accordingly, walk hand in hand. Holloway's critique of fetishism draws together the concerns of our two volumes, by urging that theories which take their object for granted, instead of asking why their content takes the form it does, reinforce the fetishism subsisting in practical life.

Notes.

1. Cf. K. Marx, *Grundrisse* (Harmondsworth, 1973), p. 109.
2. On 'new petty bourgeoisie' see S. Clarke, 'Marxism, Sociology and Poulantzas's Theory of the State', *Capital & Class, no. 2 (1977);* J. Holloway and S. Picciotto (eds), *State and Capital: A Marxist Debate* (London, 1978), Intro.

3. See E. O. Wright: 'Class Boundaries in Advanced Capitalist Societies', *New Left Review* no. 98 (London, 1976); *Classes* (London, 1985); 'What is Middle about the Middle Class?', in J. Roemer (ed.), *Analytical Marxism* (Cambridge, 1986).

4 R. Bhaskar, *Reclaiming Reality* (London, 1989); for critique see R. Gunn, 'Reclaiming Experience', *Science as Culture*, no. 11, (1991).

5 For the politics/economics separation see G. A. Cohen, *Karl Marx's Theory of History: A Defence* (Oxford, 1978); on individualism see J. Elster, *Making Sense of Marx* (Cambridge, 1985), ch. 1. Elster makes it clear that he defends individualism in a methodological sense. However, this defence of his procedure involves him in separating methodology from first-order social thought. The same separation is to be found in philosophy of science. The difficulty is that a theory/metatheory separation involves a theory/practice separation as well: see R. Gunn in this volume. Further on analytical/rational choice Marxism see Roemer (ed), *Analytical Marxism* (Cambridge 1986); A. Carling, 'Rational Choice Marxism', *New Left Review*, no. 160 (1986).

6. See J.–F. Lyotard, *The Postmodern Condition: A Report on Knowledge* (Manchester, 1984). Lyotard declares against 'metanarratives', such as teleological stories about the emergence of freedom or truth, and, in their place, argues for a plurality of 'language games' each of which entails a struggle for power. In effect he renews, in a fresh rhetorical guise, the pluralism of 1950s Anglo-Saxon political science. He replays the old theme of an 'end of ideology'. The end of ideology becomes the end of history itself. Lyotard's discussion, accordingly, remains ahistorical. Ahistorical analyses place themselves at the mercy of the history – indeed, of the metanarratives – they deplore.

7. Cf. K. Popper, *The Poverty of Historicism* (London, 1957); also his *Conjectures and Refutations* (London, 1963), ch. 16.

8. The founding text of post-empiricist philosophy of science is T. S. Kuhn, *The Structure of Scientific Revolutions* (Chicago, 1962). Cf. P. Feyerabend, *Against Method* (London, 1975); R. J. Bernstein, *Beyond Objectivism and Relativism* (Oxford, 1983); and the rehabilitation of pragmatism in R. Rorty, *Philosophy and the Mirror of Nature* (Oxford, 1980).

1

Against Historical Materialism: Marxism as a First-order Discourse

RICHARD GUNN

Marx is famous as a social theorist, that is, as one who adopts 'society' as an object of theoretical reflection. The aim of the present piece is to argue that he is neither a social theorist, nor indeed a 'theorist', in this sense. (By 'theorist', I understand one who reflects upon any object whatever, the object concerned being specified in some set of concepts, or categories, or terms.) An initial version of my point can be reported as follows: Marx offers not a social theory but a critique of social theory in the same sense as he offers a critique of philosophy (in his early writings) and a critique of political economy (commencing in 1844 and renewed in the *Grundrisse* manuscripts of 1857–8 and *Capital*). The same point can be stated by saying that Marx was not a sociologist but a critic of sociology: 'Marxist sociology' is a contradiction in terms. (By 'sociology' I understand not merely the specific discipline erected on neo-Kantian foundations in the latter half of the nineteenth century – cf. Rose 1981, ch. 1 – but any general theory of society whatever; a definition of 'general' theory is supplied below.) A further statement of my contention runs thus: *historical materialism is unmarxist*, and not just for reasons of terminology[1] or of the economic determinism to which, for instance, the formulations of Marx's 1859 Preface to his *Contribution to the Critique of Political Economy* succumb.[2]

Before proceeding to specific arguments, it is necessary to set the above contentions in context. *Politically* they lie within a broadly-construed 'autonomist' outlook; *theoretically* – and it is to their theoretical aspect that I here confine myself – they originate with an exploration of the relation between (first-order) 'theory' and (second- or higher-order) 'metatheory' in Marxist thought. Marxism refuses

1

the conventional distinction between discrete theoretical, e.g. socio-logical or scientific, and metatheoretical, e.g. philosophical or meth-odological, bodies of thought. This refusal is consequent upon the species of *reflexivity* for which Marxism declares (a theory counting as 'reflexive' when it reflects upon the validity of the categories, terms, etc. it employs). Marxism's specific requirement is that theory be *practically* reflexive (a theory counting as practically reflexive when it reflects upon the validity of its own categories *in the course of* reflecting upon its own practical situation, and *vice versa*: cf. Gunn 1987a). Specifically *practical* reflexivity undermines the conventional theory/metatheory division inasmuch as its permits first-order – practical, 'social', etc. – reflection to impinge metatheoretically, and *vice versa*: as it were, it advances simultaneously and in one and the same theoretical movement upon 'theoretical' and 'metatheoretical' fronts. It is in this way that dialectical thought comprises totalisation for Marx and, I would argue, for Hegel as well.

In other words the theory/practice unity contended for by Marx entails a theory/metatheory unity. Elsewhere (Gunn 1989a), I have followed through some of the implications of this along the axis of 'metatheory': the implications are to the effect that Marxism contains no conceptual gap which philosophy, or methodology, might fill. It contains no such gap, not because it dismisses (*á la* positivism) the questions philosophy asks but because it condemns their philosophi-cal form, i.e. their appropriation into the field of a discrete discipline. In the present article, it is with the first-order or 'theoretical' axis of theorisation that I am concerned. Once the theory/practice totalisa-tion is sprung apart, with the consequence that theory and metath-eory part company (becoming the objects of separate disciplines), not just the relation between first- and higher-order theorising but the character of each becomes affected. In the first place, they stand in relation to one another as opposite but complementary sides of the same coin, just as the state is the obverse side of the coin of civil society, and civil society the obverse of the state, for Marx. But, in the second place, conundrums break out within each of theorisation's now mutually exclusive parts.

Metatheory reserves to itself the task of validating, or criticising, theory's categories. It *must* do so, granted the premise of a theory/ metatheory severance, because otherwise first-order theory would have to validate its own categories with the consequence that vicious circularity would result.[3] But if it *does* do so then, granted the same premise, infinite regress is substituted for vicious circularity: a second-order theory requires a third-order theory for its own validation . . . and so on, without hope of cessation. The threat of just such a regress echoes through the history of philosophy, from the 'Third Man' argument in Plato (cf. Cornford 1939) to disputation over the argumentation for the validity of categories in Kant.

First-order theory devotes itself to the task of explicating, by whatever means, a cognitive object specified in metatheoretical terms. First of all metatheory defines the object of theory, and then theory sets to work. To be sure, these days, matters are not quite so straightforward. Metatheory is inclined to adopt an *underlabourer* role, as in Kuhn (1962), Althusser (1976) or Bhaskar (Bhaskar *et al.* 1988; Bhaskar 1989). In Kuhn, the underlabourer position takes the form of assimilating philosophy of science to history of science; in Althusser, ambiguously,[4] it takes the form of a historicisation of philosophy; in Bhaskar the underlabourer function appears in the shape of transcendental argument from the practice of science to the reality of the causal structures science reveals.[5] (So called 'post-empiricist' philosophy of science is largely a restoration of such philosophy to the underlabourer status bestowed upon it by Locke, three hundred years ago: in the 1980s and 1990s, humility appears to be all the rage.) But this shift to underlabouring achieves nothing (for instance Bhaskar remains more hubristic than humble in the passages where he urges that Critical Realism, as a metatheory, is able to set out of bounds such sciences as do not acknowledge the reality of causal structures). Indeed the hubristic/humble dichotomy fails to capture the complicity – the two-sides-of-the-same-coin aspect – which erupts between theory and metatheory once they are construed as reciprocally distinct. *Either* the underlabourer manoeuvre amounts to sheer relativism, simply taking the practices of first-order theory on trust, *or* it amounts to the proposal that theory sets the problems which it is up to metatheory to solve. In other words it requires that 'theory' wait upon metatheory's solutions (just as metatheory perforce waits upon theory's questions) and *vice versa*: popular consciousness awaits the solution of philosophical problems by scientific means. 'Once the metatheorists have got it sorted out we can start working'; 'once the theorists get it established, finally, we can draw conclusions': each movement in this sort of argument takes in the other's washing. Respectively, either infinite regress or vicious circularity are the consequence of arguments pitched in this way.

For the purposes of the present paper, it is the theoretical as distinct from the metatheoretical consequences which concern us. And they can be summed up as follows: any 'theory' is a *theory of* . . . That *of which* theory is a theory is specifiable only metatheoretically, even if the interplay of theory with metatheory (announced by the underlabourer strategy) comes into play. A 'theory of' requires an object which is categorially specific; theory of a categorially specific object is 'theory of'. These days, we tend to think of all theory whatever as 'theory of', that is, we tend to assimilate theorising *per se* to the standpoint of what Hegel (e.g. 1977, Intro.) terms 'consciousness'; the characteristic of 'consciousness' in its Hegelian meaning being that it construes its object – and thereby *constitutes it as an object* – as something which stands over against the subject who lays claim to know it. Consciousness, in short, and theory *qua* 'theory of', runs the risk of *reifying* its object. This point is of such fundamental importance that I permit myself elaboration of it at greater length.

Because 'theory of', or 'consciousness', tends to reify its object, Hegel links it with the practical statute of an alienated world. Twentieth-century phenomenology reports the same thing: according to Merleau-Ponty (1962), consciousness tends to overshoot itself from first-person into third-person views. If, then, theory *per se* appears to us as 'theory of', and if consciousness *per se* strikes us as thus overshooting, this, on Hegelian and Marxian premises, can be attributed to the estrangement of the world in which, as so to say theoretico-practical beings, we live. Reification within cognition reports alienation within practice, and *vice versa*. So dense is this alienation that we find it difficult to imagine what theory might be, if it ceased to be 'theory of'. Only on the margins of epistemological discourse is the *soi disant* presentation of theory as 'theory of' challenged. For instance, a refrain in Frankfurt School Critical Theory is to the effect that truth and falsity pertain not just to theorising about facts but to the facts themselves. Or again: 'In a world which *really is topsy-turvy*, the true is a moment of the false' (Debord 1987, para. 9). In fact, as the previous citations demonstrate, construal of theory as anything other than the theory *of something* (of concepts, of societies, of nature: but all these terms are problematic) is likely to find its only cultural resource in paradox.

To see that something more than wilful paradox is involved we have to examine the notion of 'theory of' more closely. A 'theory of' anything whatever is what I shall term a *general* theory. A general theory is one which addresses *all of its object and only its object* (this being the 'extensive' condition of general theory) while at the same time (this being its 'intensive' condition) it *neither overdetermines nor underdetermines its object*. For instance, a general theory of society (extensive condition) would have to be competent to address all societies but only societies, this latter part of the condition excluding theological and naturalistic conceptual schemes. (To be sure, such a theory is competent to reflect on e.g. society's relation to nature, as does historical materialism, and even to make this relation pivotal, but only on condition that it construe the society/ nature relation as a social relation for its part: nature is a 'societal category', as Lukács reports (1971).) Further, a general theory of society (intensive condition) must neither underdetermine its object, by for example introducing theological or naturalistic explanations, nor overdetermine it by privileging some aspect of social existence, such as the economic or the cultural: it must address the 'society effect' in Althusser's (1970) phrase. (To be sure, some aspect of society may turn out to be *de facto* privileged, but no such privilege can be established on methodological grounds.) To specify *both* extensive and intensive conditions for general theory may seem needless, inasmuch as a *successful* general theory is one in which extensive and intensive conditions are conjointly met: if 'society', to resume our example, is theoretically neither under- nor overdetermined then all of society and only society will be theory's object, and *vice versa*. But it makes sense to distinguish the conditions because each of them plays a different role in general theory's constitution. The extensive condition supplies the (in a hermeneutical sense) pre-judgement or preconception of general theory's object: in the above instance, we all know informally and approximately what 'society' means. Hence the extensive condition points to something indispensable, namely general theory's starting point, but if *only* the extensive condition were in operation general theory would be consigned to the wholly implausible domain of reporting brute, i.e. category-neutral, facts. (And whatever its aporias it has almost always resisted entry into this domain.) The intensive condition of general theory, by contrast, comes into play once extensivity has posed the problems: if we are to know for instance *what society is* – if we are neither to overdetermine nor underdetermine it in our knowledge of it – then we are compelled to address the issue of the terms and concepts and categories by which it may be known. Intensivity reports a second step in the constitution of general theory, albeit a step which involves (how well are the initial questions posed?) retrospective interrogation of the first. And, in fact, it is the intensive condition which has logical priority. For unless adequate categories can be enunciated for the prospective resolution of a problem the problem falls. This priority is reflected in my definition of general theory, which construed it as in effect generality *with respect to some object*. (A general theory of a single molecule, on this definition, would not be impossible.) The 'with respect to' clause foregrounds intensivity, i.e. the metatheoretical issue of determining the categories by which this or that object may *be* an object or in other words something potentially known.

General theory is accordingly 'theory of', and *vice versa*: first-order theory becomes general theory once – by virtue of a theory/ practice disunity, which is to say by virtue of alienation – its metatheoretical dimension is prised away.

Our specification of 'theory of' as general theory brings something important to light: firstly, *the relation between universal and particular* characteristic within purely first-order thought. Inasmuch as intensivity takes priority over extensivity the universal takes priority over the particular: the *only possible* objects of 'theory of' are those which can be brought within the ambit of universal categories of some kind. To be sure, these universal categories may be known prior to encounter with the particulars (Kant's 'determinate judgement', predominant in his first two critiques) or they may be generated in and through contact with the particulars (as in 'underlabourer' approaches, together with the 'reflective' judgement of Kant's third critique – but this latter already begins to break with general theory *per se*). What remains suspect in both approaches, the determinate and the reflective, is the view that particulars which cannot be subsumed beneath universals are not worthwhile objects of cognition at all. Either they are nonexistent or trivial or revolting: Plato for instance (cf. Cornford 1939) averred that there were no Forms, or universal Ideas, of 'hair' (pubic hair?) and 'dirt'. Here, in fact, we are in the presence of a central motif of Western thought, which has always assumed theory to be general theory ('theory of') ever since Socrates began to speak. How can we know particulars – how can we allocate to them 'categorial rights' – without subordinating them to universals? Kant, for instance, in his first critique went so far as to contend that twelve and only twelve categories were the 'lenses' through which objectivity might be uniquely seen. The problem of relating these universals to particulars surfaces in his discussion of 'schematism' and the related notion of 'productive imagination' in the *Critique of Judgement* (cf. Walsh 1975): which takes logical priority? Schemata or productive imagination? The first critique or the third? Kant scholars have never been able to decide.

Once again, we are returned to the margins of epistemological thought should we seek a rehabilitation of the particular. *Nominalism* foundered (a) on the problem of induction – which is in effect Kant's entitlement out of Hume – and (b) on the self-evident circumstance that all knowledge of particulars, for its part, comes through universal categories. Frankfurt School theory accordingly reports that any cognition refusing the murder and dismissal of particulars involves thought (which as such deals in universal categories) thinking against itself (Hegel had already equated sheerly universalistic 'understanding' with murder in his *Phenomenology of Spirit*). Adorno (1973) winds up the paradox to the extent of declaring particularity itself to be a universal category. The twentieth century has seen a rehabilitation of particulars *surrealistically* in, for instance, the doctrine of truth as a 'constellation' of particulars rather than as an induction from them in Walter Benjamin's introduction to *The Origins of German Tragic Drama* (the idea here being that particulars are 'monads', containing the universal within themselves) and in the early writings of Bataille (1985). This rehabilitation remains untheorised, however. To demonstrate that particulars are already *qua* particular universal leaves it an open question how universals *qua* universals contain particularity in themselves.

The second point of importance arising from our characterising of 'theory of' as general theory concerns the relation between genus and species which general theory entails. This relation is the familiar tree-diagram one of individuals falling beneath species which fall beneath genera in their turn. In the event, that such relations obtain (general-theoretically) is merely an inference from the point concerning universals/particulars discussed above. But the point can be unfolded from the notion of general theory *per se*.

General theory counts as 'general' to the extent that it can marshal a field of particulars into an 'object' specified, intensively, in some set of categorial terms. Only then can theory count as theory of. But, precisely insofar as this marshalling is successful, further extensive questions (above and beyond specific intensive programmes for systematising problems and generating prospective resolutions) break out. Can the data supplied by *this* general theory be drawn under the wing of a general theory of, so to say, a still more general kind? Can the constitution-movement from extensivity to intensivity become the starting point for a further movement of the same sort? Such questions derive their logical force from the circumstance that no knowledge is category-neutral (there are no 'brute facts') so that *data as such*, even when it is generated as theory's outcome, can never figure purely as a result; it must also count as a fresh starting point for reflection even should all concerned accept it as true. *Data* raises new questions and *true* data raises questions of a generic as opposed to a specific kind. Problematic data passes us up the metaladder of infinite regress; well ascertained data passes us up the ladder which ascends from species to genus, and so on. (Equally it passes us *down* the ladder: can species not be individualised and particularised *ad infinitum*, or, at any rate, increasingly?) So to speak, 'theory of' flattens out the vertical ascent through metatheories into a one-dimensional ascent obtaining within first-order theorising alone. Foucault's *The Order of Things* discusses the eighteenth-century rendition of this point eloquently. To be sure, the nineteenth and twentieth centuries have cut away from any such Aristotelian schemes. But to the extent that these severances are successful they place a question-mark not just against a local and cultural Aristotelianism but against the notion of general theory itself. The logical structure of general theory *allows* it to sit at mid-point between individual and genus. Once this mid-point is deconstructed the overall structure of general theory totters or, better, falls.

The above has attempted a sketch of the framework of a first-order theory progressing under the sign of a metatheory separated from it. The charge brought against such theory, quite apart from logical ones, is to the effect that it swims in alienated waters: the theory/practice split. I have now, so to speak, established legitimacy for the currency which the remainder of my paper will expend in a more or less polemical way. In doing so, I shall outline an alternative to theory as general theory. Establishing such an outline and carrying through theoretical reflection within it I take to be the unique (the unsurpassed) achievement of Marx.

What has to be dispensed with? In the first place, and in my earlier definition, sociology. Sociology (as any general theory of society whatever, whether voluntarist or determinist, individualist or structuralist, Weberian or Durkheimian, 'bourgeois' or 'Marxist') perishes once the notion of general theory is overthrown. Marx *cannot be* a social theorist – again, in my earlier definition – if he takes his thesis of a unity between theory and practice seriously. In fact he *is not* a social theorist inasmuch as he declares ('Theses on Feuerbach', VIII) that 'all social life is *essentially* practical' and, to the same effect in the *1844 Manuscripts*, that alienated labour is the key to the institution of private property rather than the converse. The emphasis of such categories is on action, or practice, in contradistinction to social structures: if you want to understand 'society', he seems to be saying, do not look at society *per se* but at the practice from which (currently, and under the sign of alienation) society results. The temptation is of course to read such passages in the sense of advocacy for an action-oriented *sociology*. But, in the light of the above discussion, we can now read them better: they announce (broadly speaking *avant la lettre*) a critique of sociology as such. Even an action-oriented sociology – and quite apart from the individualist overtones which such sociology has generally carried – remains a theory of society: and it is these two complementary things, a 'theory of' and a 'society', which Marx energetically contends against. 'Theories of' and 'societies' subsist only once the theory/practice unity has fallen apart.

As mentioned earlier, these reflections overturn everything coming forward in the guise of a Marxist sociology, however conjunctural such a sociology may be, for, ultimately, 'generality' was defined intensively. Not only is small-scale analysis no recourse, but small-scale relates to large-scale as does species to genus: each legitimates the other. For instance, the fordist/post-fordist debate belongs within the genre 'sociologies of transition', sociology of transition having been more or less the definition of sociology *per se* ever since the author of *The Protestant Ethic and the Spirit of Capitalism* took up his pen. Or again Gramsci, who projects conjunctural analysis paradigmatically, is as much dependent on élite-theories as on Marxism (Bellamy 1987). Finally, the offer of Rational Choice theory to reconstruct a macropolitics of class out of a micropolitics of individual decision (Elster 1985; Roemer 1986) turns upon general questions at the moment when it brings the notion of equilibria into play. The point is not that, in each of these conjunctural schemes, an un-Marxist general theory is invoked; rather, it is to the effect that the general theories which are invoked *have* to be un-Marxist, for Marxism, as construed above, contends against a general theory (e.g. a general *theory of* society) of whatever kind.

Secondly, historical materialism has to be dispensed with. Historical materialism, socalled, is the nearest approach to a general theory of society which Marx's writings – notably *The German Ideology* Part One and the 1859 Preface – contain. The 'general' character of historical materialism can best be illustrated by reference to the problems to which it has given rise. *Extensively*, there awaits a Marxist reckoning with the Asiatic mode of production (Wittfogel 1953; Lichtheim 1963; cf. Hobsbawm 1964, Intro.), a reckoning made the more politically urgent by Bahro's (1979) indication of the Asiatic mode of production as the site where a crisis of Marxism might be expected to break out. Debates on the status of an Asiatic mode refer, of course, within a historical materialistic framework, to history's beginnings; no less logically powerful extensive considerations break out at history's end. An entire school of Marxism (Lukács 1971, fifth essay; Marcuse 1941; Horkheimer 1972, p. 229; Sartre n.d., p. 34) avers that historical determinism is all very well as a report of capitalism, which is deterministic *de facto*, but no sure guide to the emancipatory existence towards which the antagonistic relation of labour to capital clears the way. This latter school abuts on to *intensive* considerations even while foregrounding extensive ones: the problems here concern the overt economic determinism of Marx's 1859 Preface. For instance, according to Lukács (1971, p. 27): 'It is not the primacy of economic motives in historical explanation that constitutes the decisive difference between Marxism and bourgeois thought, but the point of view of totality.' To be sure, formulations like 'motives' and 'points of view' still remain too close to the Weberian paradigm in which Lukács had been educated; methodological individualism (the obverse, Weberian, side of the coin of structuralism) remains their norm. But it is clear enough that Lukács is saying – along with, later, Marcuse, Horkheimer, Sartre and indeed Gramsci – that economic determinism (a) underdetermines social existence by assimilating it to the sort of 'society of beavers' alluded to by Marx in the *1844 Manuscripts*, while at the same time (b) overdetermining it by bracketing off, as superstructural, the entirety of the social realm 'in which men become conscious of . . . conflict and fight it out' (Marx 1971, p. 21). The equivocation within historical materialism between totalising and causalist perspectives (see fn. 12, below) throws into relief its problematic character along the intensive axis of generalising thought.

In other words historical materialism falls victim to the Marxist critique (deriving from the thesis of a theory/practice unity) of theory as 'theory of'. The Leninist and Engelsian idea of *Capital* as a specific application[6] of a generic historical materialist conception goes by the board (along with genus/species relations themselves). So, too, do contemporary fordist/post-fordist sociologies – and for reasons quite

other than those raised in debates concerning their Marxist provenance. Their historical materialist provenance can be allowed but,
still, they fail. The critique of historical materialism here raised
addresses less its content than (as a general theory, intensively
specified) its form. Debates within, and concerning, historical
materialism, have of course been numerous. For example there is
the recurrent question of whether historical materialism is totalising
or causalist; the connected question of whether an economic 'base'
can be conceptualised independently of a political and legal and
ideological 'superstructure';[7] and a dispute in regard to the question
of whether 'forces' of production have priority over 'relations' of
production or *vice versa*.[8] The question of the precise definition of
forces and relations of production remains no less controversial.[9]
None of these debates addresses the question of historical materialism's status, or form. Hence, from the present point of view, they
count as secondary. Perhaps the only sensible word which has been
spoken in the course of them is to the effect that Marx's 1859 Preface
omits (in the course of describing the 'guiding thread' of his studies)
all mention of class, whereas the opening sentence of the *Communist
Manifesto* characterises this thread exclusively as the practice of class
struggle.

It was mentioned earlier that historical materialism belongs
amongst the least original departments of Marx's thought. Spotting
'anticipations' of historical materialism has become something of an
academic hobby (e.g. Pascal 1938; Skinner 1965; Meek 1976). To
Comninel (1987; and cf. Gerstenberger's contribution in Volume
One) we are indebted for the suggestion that, whereas in Marx's
writings political economy is processed through critical mills, he
takes his conception of history – and thereby of the 'society effect' –
more or less on trust. Anticipations can be readily discovered only
because, in historical materialism, there is nothing new. Comninel
foregrounds *struggle* as the crucial historical materialist issue, and his
work contains a magnificent diagnosis of the teleologism inherent in
sociologies of transition (cf. Bonefeld 1987 for an argument to the
same effect as regards fordism/post-fordism), but still it is a *historical
materialism of* struggle which Comninel wants. My own contention is
that struggle detonates (explodes and implodes) the boundaries of
historical materialism, and points beyond its confines, inasmuch as a
'theory of' struggle – a general theory with struggle as its subject-
matter – is a contradiction in terms. The argument supporting this
contention has been presented above. *Either* struggle *or* the reifica-

tions of general theory: between them there lies no middle, or third, way.[10]

This is not to say that all of the arguments within and concerning historical materialism are equally vacuous (or valid). Some of the arguments threaten a break with historical materialism's form whereas others do not. Those which do are the arguments premised upon totality, for they bring the question of the relation of universality to particularity into play. The genus/species schema according to which particulars (in a telling phrase) fall beneath universals is problematised by a conception of totality which, however simple or 'complex' (cf. Althusser 1969, 6; 1970, ch. 5), records the presence of the universal within the particular and *vice versa*. When Sartre (n.d.; 1976) polemicises against Lukács, rightfully seeking to stress the activity of totalisation as against the passivity of 'totality', this amounts to a minor though welcome correction, or in other words to an in-house debate: totality signals nothing other than the movement of contradiction (cf. Hegel's *Jena Logic*: Hegel 1986, p. 35ff.). Once particularity receives its due, generality – or rather *pure* generality, as defined earlier – detonates. Lukács *et al.*, therefore, move out of the orbit of general theory not just extensively but intensively as well. The question they raise is that of the historical specificity of *soi disant* general, or universal, categories: Marx's interrogation of fetishism, or reification (cf. Goldmann 1977), problematises nothing else.

I permit myself, at this point, a parenthesis. If anything whatever in the above chain of argument is found valid, or intriguing, then there is no option but to revise a conception of *Marx's intellectual development* which has become a virtual orthodoxy in its own right. This conception divides Marx up as between his 'early' and his 'mature' writings, weighting the one against the other equally or unequally as the case may be. The most extreme version of this periodisation is of course Althusser's: Althusser (1969; 1970; 1976) urges that an 'epistemological break' erupts within Marx round about 1845. Prior to 1845 Marx was an ideological humanist; subsequent to 1845 Marx began to lay down the foundations for a historical materialist science (whose outlines Althusser finds himself unable to make clear).[11] David McLellan's numerous commentaries on Marx (their *summa* being McLellan 1973) proffer a watered-down version of the same thesis: Marx (*pace* Althusser) did not stop being a humanist in 1845 but none the less the historical materialism adumbrated in the 'Theses on Feuerbach' and consolidated, at least provisionally, in *The German Ideology* represented a profound step forward in his thought. The implication of the argument presented in this essay is that it amounted to no such a thing. The historical materialism of 1845 (and remembered, inaccurately, in 1859)[12] signals less a theoretical advance than an 'infantile disorder' (Lenin) which beset the critique of political economy, and of society, in its early days. The true line of continuity extends from 1844, when the critique of political economy was first enunciated, albeit within a moralistic framework of competition, to 1857 when the 'Introduction', subsequently replaced by the 1859 Preface, was laid out in draft. What is astonishing in Marx is the single-minded continuity of his class-oriented writings from first to last. Within the line of this continuity historical materialism figures as something which the 1980s have familiarised us with, namely, a blip. It is as though Marx, having announced the critique of political economy and of society, faltered and fell back upon a *social theory* as the underpinning for his views. The 1859 Preface, whatever its distortions, reports this circumstance honestly: it is at the end of this Preface, Marx tells us, and not in the course of it, that we stand at 'the entrance to science' (Marx 1971, p. 83). In the very last decade of his life we find Marx fighting the same battle no less vigorously: the 'Notes on Wagner' report 'value' to be quite other than a 'concept', i.e. a genus beneath which species of commodity production might be ranged.

But what might Marxism be, *minus* historical materialism? The overall answer to this question is implicit in what has gone before. Marxism announces struggle in place of stasis, activity (or practice) in place of passivity, subjectivity in place of substance, particularity in-and-through universality in place of universality (and its genus/ species differentiation) alone, etc. In other words, we can initially identify Marxism's concerns by means of privileging the first of each of the terms announced in a list such as this over each of the second terms. Marxism, along its first-order axis, approximates less to an action-oriented sociology – and this has nothing at all to do with the typically methodological individualism of such sociologies, against which Marxism also declares (1976 p. 280) – than to a critique of sociology in the name of action.[13] This is not to say that the second term in each of the above-mentioned pairs is unreal, or trivial, according to Marx. When social relations between commodity-producers appear, to them, as 'social relations between things' they appear as 'what they are' (Marx 1976, p. 166; cf. Geras 1972). We can fall beneath policemen's truncheons – a social reality – just as easily as we can fall beneath universals or over physical objects in the dark. Stasis *exists*, in the Marxist conception, but it exists as struggle subsisting alienatedly, i.e. *in the mode of being denied*. Structure exists as practice, but in the same mode (Gunn 1987b). The second of each of the pairs just reported – and the list could be extended – signals a parasitism, i.e. a dislocation of theoretical objects from 'the sheer unrest of life' (Hegel 1977, p. 27). Non-alienated being, reports Hegel (1977, p. 265), 'no longer places its *world* and its *ground* outside of itself'. Marx's polemics on alienation, derived from Hegel, say the very same thing. General theory, even in a historical materialist guise, disrupts everything. A political practice guided by a general theory consolidates rather than decon-structs – its deconstruction is only a polite one – the demarcations of a territory through which it charts a path. All this is, I take it, on the basis of the above discussion, perfectly clear.

What is clear is a specification of the terrain of Marxist theorising spelt out in terms of what such theorising is not. This, arguably, is sufficient given the sonority of Marx's silence – as if it were the tolling of dampened bells – which rings out across his discourse concerning *this* world at the very points where a discourse concerning an *emancipated* world seems to be announced (Gunn 1985). Were this article to be authentically Marxist it would end here. Commu-nism, for instance, is nowhere reported by Marx as a form of 'society' but only as (in *The German Ideology* and *The Civil War in France*) the real – or, better, actual[14] – movement of the working class. Communism, thus, *already exists* – 'existence' being here understood in the sense of ek-stasis *á la* Sartre or Ernst Bloch. Existence, in Bloch's phraseology, is always and already existence not-yet. Marxism 'is a theory not of oppression but of the contradictions of oppression' (Holloway 1989); and these contradictions move. This said, however, I shall attempt an account of what Marxist theorising *is*. Some more polemical asides are, after all, called for. To anyone who has followed through my argumentation and endorsed it the remainder of my paper is, strictly speaking, *de trop*. Stop reading here.

The more emphatic or positive account of Marxist theorising turns upon a deepening of the notion of practical reflexivity, signalled above. This is what Marx undertakes in his (abandoned and thus silent) 1857 Introduction (Marx 1973, pp. 83–111). This text is one of the most disputed within Marxism – it was published by Kautsky, cited by Lukács against Kautsky, declared to be structuralist by Althusser and deemed humanist and totalising by McLellan – and, no less, it reports the critique of Marx's historical materialist *alter ego* as undertaken by Marx himself. Everything here flows. The focus of the text is upon an issue which may seem rebarbative, or academic: viz. abstraction. Depending upon how we construe abstraction, Marx seems to be saying, turns our politics on the right or the wrong side of the class struggle (Negri 1984, Lesson One; 1988, ch. 4). In other words, seemingly esoteric conceptual points can carry along with them directly exoteric political connotations. That they *must necessarily do so* is the force of Marx's 1857 critique. How we think is not innocent. Whoever tries to rise from the particular to the the generic, or universal, – whoever remains faithful to the programme of what used to be called 'inductivism' – remains (Marx 1973, pp. 100–1) trammelled within a world where false facts may be true and true facts false. A 'chaotic representation [*Vorstellung*]' is the result: in Hegelian terminology *Begriff* lies, still, ahead. The way to transmute *Vorstellung* into *Begriff* is, says Marx (1973), to ascend not from the concrete and particular to the abstract but from the abstract to the concrete. Here of course the danger of idealism (in the sense of *a priori* and practice-disconnected thinking) breaks out. Marx declares against such thinking in the Introduction (pp. 101–2) while sometimes appearing to lapse into it (Preface to the First Edition of *Capital*: Marx 1976). The question arises: what systematic defences against such a course – as it were, deductivism in place of inductivism – did he command? The remainder of the 1857 Introduction makes everything clear.

Inductivism is 'abstraction from'. It offers to bring particulars within the ambit of universals and the critique of it (from Hume to Popper) declares, rightfully, that it cannot. *Deductivism* is 'abstraction into'; its standpoint is that of Kant's 'determinate judgement'; its offer is to subsume all possible particulars under universals of some kind. The rightful condemnation of deductivism (from the Sophists whom Socrates/Plato targeted to Adorno/Benjamin and Bataille) is to the effect that particulars are recalcitrant: against any conceivable system of classification the particulars, especially when they are human, tend to fight back. Thus both 'abstraction from' and 'abstraction into' – let me call any conception of abstraction which depends upon the one or other of these strategies *empiricist* abstraction – fail. If Marx's (silent) 1857 Introduction is to support its case then it has to develop a conception of abstraction of neither an inductivist nor a deductivist sort. That indeed the Introduction *is* silent testifies to Marx's consistency regarding the thesis of the unity of theory and practice: 'a general introduction, which I had drafted, is omitted since . . . *it seems to me confusing to anticipate results which still have to be substantiated*' (Marx 1971: emphasis added); the substantiation, I take it, is (as always in Marx) of a practical and political kind. In other words Marx's 1857 appeal to the non-inductivist and non-deductivist conception of abstraction adumbrated in his 1857 Introduction – consigned to the silence which builds communism – has the status of a *wager*[15] upon the outcome of the actually present struggle of class. 'Human anatomy contains a key to the anatomy of the ape' (Marx 1973, p. 105): this, I take it – and I take it so because the above has shown that Marx *cannot be* a general theorist – is a declaration in favour not of some sort of Aristotelian teleologism but of the notion that abstractions prefigure a radically open future, much in Bloch's (1986) sense. Marx announces a *temporality of abstractions*, which both 'abstraction from' and 'abstraction into' deny. Apes do not tend, teleologically, to become humans; but humans, because their practice is premised upon a projection of a novel future, can make sense of apes – and of their own past. Abstractions for Marx exist in time; and by the same token, universal though they be, they exist as particulars for their part. Instead of universalistic thought having to think against itself (Adorno) in order to capture particulars, we have the particularistic – historical social, practical – existence of universalistic thought: in other words, totalisation (Marx 1973, pp. 100–1).

The example of a universalistic category, construed in this way, which Marx offers in the 1857 Introduction, is that of *labour*. Labour as an 'abstraction from' or an 'abstraction into' obtains, to be sure, in 'all epochs'; however, labour as an abstraction existing – abstractly – *in practice* obtains only once capitalist social relations (commodity production so general and intensive that even labour-power subsists only as a commodity) comes to the fore (Marx 1973, p. 105). Anyone who sets out to sell the use of their labour-power in the labour market discovers immediately the truth of the proposition that *labour is abstract*. Hegel's contention that the abstract understanding is 'murderous' is here replayed in a practical mode (as a report on the deaths of workers).

This said, it has to be confessed that Marx's example of 'labour' is ill-chosen. Perhaps *any* example would be ill-chosen, inasmuch as examples *per se* inhere, as a mode of rhetoric, in the genus/species conception of abstraction – the conception from which, in 1857, Marx is breaking free. Throughout his life Marx declared against the political economists who understood labour in a generic, and thus a general, way. A more telling example would have been not labour, but the capital/labour interrelation: only within this interrelation (a relation of struggle) can labour *as an abstraction subsisting not just in theory but in practice* appear. 'Work [itself a problematic category] which is liberated is liberation from work' (Negri 1984, p. 165): communism, this liberation, already ex-sists. In accordance with the above argument, the example of the capital/labour relation, instead of labour, would have better illustrated the deconstruction of empiricist abstraction in Marx's (post-1857) thought.

What might a thought renouncing empiricist abstraction look like? Surely, it might look like the *Grundrisse* and *Capital*: Marx was the first-ever (and only) social theorist to make a point of principle out of dealing only in terms of abstractions which exist in practice. Marx thus historicises, or totalises, his own thought in a radical way. I stress the word *only* in case this reflection appears banal, for after all (from McLellan and humanism through to Althusser and structuralism; and beyond) no-one has ever denied that a thesis concerning the unity of theory and practice controls Marx's thought.

Of course everything turns on how this thesis is understood. Above, I have suggested that it is to be understood in the sense of *practical reflexivity*. This allows Marx to understand abstractions not merely as socially situated (as ideological forms) but as socially real. *All* social life, abstraction included, is essentially practical. To be sure there can be socially unreal – or, better, unactual – abstractions; but only practically reflexive thought can illumine the existence of abstractions (labour, class, value, etc.) which *are* real; and moreover practical reflexivity *requires* the interrogation of abstractions with regard to the reality which is, or is not, theirs. In other words it requires, and renders possible, a distinction between empiricist abstraction and what might be termed substantive or *determinate* abstraction (Gunn 1987b) – the latter being such abstraction as inheres not just in theory but in practice as well.

The continuity of Marx's work lies, thus, along a line which points directly from the early writings (up to and including the 'Theses on Feuerbach'), wherein the thesis of a theory/practice unity is articulated, to the 1857 Introduction, where the notion of determinate abstraction first appears. *The German Ideology*, which announces historical materialism, and more especially the 1859 Preface, which misremembers *The German Ideology*, count only as digressions along the way.

Upon this notion of determinate abstraction much depends. Empiricist abstraction is purely first-order abstraction: to it belongs the distinction between species and genus and the problematising of universals over particulars. Empiricist abstraction is the guiding thread of the 1859 Preface as distinct from the 1857 Introduction. This reflection goes some way to explain the *causalist* tone of Marx's 1859 remarks. To be sure, not all analysis which privileges universals over particulars needs to be causalist (all metaphysical schemes of an *a priori* kind from Plato to Kant would stand as exceptions to such a rule) but, no less surely, causalist schemes are consonant with privilege of just this kind: causalism's demotion of particulars lies in the circumstance that, as a condition of causal explanation's intelligibility, the same cause must be seen as generating the same effects *everywhere* and *everywhere*. Particulars necessarily figure only as instances, falling beneath universals that (*qua* universal) are indifferent to their diversity: causalism and the genus/species/particular differentiation go hand in hand. And once universal categories are thus privileged along the theoretical or first-order axis of theorisation, infinite regress along the metatheoretical or higher-order axis begins; in other words, the axes separate. So far as Marx is concerned, notice that we are here taking exception not merely to the specifically *economic* determinism of the 1859 Preface's content but to the causalism (the weakly or strongly deterministic character) of its form. *All* causalisms, even action-oriented ones and 'causalities of freedom',[16] stand condemned in the light of the above remarks. Moreover, inasmuch as causalism privileges universals *and thus* relies on a genus/species division, we discover grounds for taking exception to the 1859 Preface's specification of modes of production ('mode of production' being understood here universally and generically) as 'Asiatic, ancient, feudal, modern bourgeois' and so forth (Marx 1971, p. 21).[17] Precisely such a *sociology* is un-Marxist, and is so quite regardless of whether (cf. Althusser 1970; 1971, p. 98) these specific modes of production be construed diachronically – as 'epochs' – or synchronically, or as the unity of the two. What is objectionable is the form of the specification itself (cf. Comninel 1987; Gerstenberger's contribution in Volume One). Moreover it is objectionable whether or not there is a plurality of ways forward, or only a single way, along history's path (Hobsbawm 1964).

Empiricist abstraction is the mode of discourse of general theory, because inductivism and deductivism (and transcendental deduction) await, alike, metatheoretical validation of the universal, categorial lenses through which particulars are to be seen. Once universals can be seen as, themselves, particulars – and *vice versa* – everything changes: the educator (*qua* metatheorist a theorist and *vice versa*) can become educated. Precisely the notions of *self*-education and self-emancipation which Marx ascribes to the proletariat[18] depend on the theory/metatheory interrelation being seen in this way. Stated otherwise: *determinate* abstraction – abstraction capable of practical and particular existence – is the mode of discourse which breaks with general theory; and which in doing so simultaneously clears a way towards critical theory and critical practice.

Still a further formulation of the same point runs as follows: *totalising* theory breaks not just with causalism's discrimination between areas, elements, instances, practices or factors within social life but with the general-theory format (as outlined earlier) upon which causalism relies. Universals already stand contaminated with particularity if, between universality and particularity, an internal relation obtains. Species and particulars already *fall beneath* universals – 'murderously', as Hegel has it – once an external relation between them is elevated to a principle of thought. Form and content equate with metatheory and theory, respectively, thus fomenting all the difficulties of vicious circularity and infinite regress considered earlier, once the totalising contention that universal categories can have concrete ('actual') existence is denied. Marx's pivotal insight, that which allows us to read the *Grundrisse* and *Capital* as spectra of original glories, is to the effect that a totalisation of theory and practice, in the sense of practical reflexivity, is the key to totalisation *per se*. Even to say that theory and practice are internally related – and not just causally related, as in bourgeois 'sociologies of knowledge' – is too weak. Only as a first approximation might we say that theory and practice 'interact'. Just as there is nothing in human self-consciousness, or self-reflection, which is asocial (cf. Hegel 1977, pp. 110–11), so, and by the same token, there is nothing in even the most rarified regions of theory – cf. Marx on 'science' in the *1844 Manuscripts* – whose constitution-principle lies outwith some practical field. In *The Poverty of Philosophy* Marx condemns Proudhon for relying upon abstractions. This condemnation is all too easily misunderstood. What Marx holds against Proudhon is less the wish-fulfillment and unreality of his abstractions than their all-too-real, and conformist, subsistence – their subsistence-in-practice – within a bourgeois world. In short, Proudhon, according to Marx, failed to understand his abstractions (failed practically to reflect upon them) as determinate. To be sure, in 1847, this dimension of Marx's Proudhon-critique remained underdeveloped. But in the parallel passage in the *Grundrisse* (Marx 1973, pp. 248–9) everything is made clear.

Only totalising theory can interrogate the status of abstractions sufficiently vigorously. *Not all* totalising thought enunciates an interrogation that goes sufficiently deep. Practical reflexivity, determinate abstraction and a *rigorous* totalisation belong together: this is one way of stating the positive conclusion which our argument has so far reached.

But of course it is only one way: I take it to *have been proved* by the above remarks, however sketchily, that general social theory far from being a pillar of Marx's thought has no place within it. Practical reflexivity, and its attendant resistance to a theory/metatheory severance, turns general theory into a death-rattle. General *social* theory, for instance historical materialism, is – because once any social theory is announced the question of practical reflexivity appears as all but inescapable – precisely the rattlesnake's tail. Marx catches this tail, and shakes it: sometimes (1859) the snake bites; at other times (1857) its back is broken. Setting every appeal to humanism aside, we have to be able to distinguish the moments where poison enters into Marx's thought.

Sociological Marxism – a Marxism depending on empiricist rather than determinate abstraction – is a contradiction in terms. Historical materialism likewise, by virtue of its general-theoretical form.[19] These conclusions, so far as I can tell, stand confronted by only a single objection: if we throw overboard empiricist abstraction what might the first-order aspect of Marxist theory be? Do we not ascend into the heavens of theory as an *a priori* and 'independent realm' (cf. Marx 1975, V, p. 447) in the same movement as the anchor-line of empiricism is cut?

The reply to an objection along these lines turns upon an understanding of *form*. Abstractions can exist concretely and in practice; universals can exist *qua* particular and *vice versa;* the concrete can exist as abstract (as in the example of labour under capital, i.e. value-producing labour, according to Marx). That is to say, the *form* of something can be construed as its *mode of existence*, whether this 'something' be abstract or concrete and whether (respectively) its form be concrete or abstract. The notion of 'mode of existence' is, here, the crucial one. It links determinate abstraction to practice (cf. Gunn 1987b; 1989a). Every 'is', so to say, has a concomitant 'how';

and is dependent upon it. Saints can exist only in saintly fashion. Saintly action is the condition of the being of saints. Adjectives (as in 'the thing with many properties': cf. Hegel 1977, ch. 2) shift over into adverbs, so that action is construed as prior to thinghood, 'saintly existence' making sense only if 'existence' is read as ex-sistence or ek-stasis or ecstacy, i.e. in an active way. In the same fashion 'appearance' can be understood as the mode of existence of 'reality' (Hegel, *Encyclopaedia*, para. 131) solely on condition that appearance is understood as appear*ing* and on the further condition that 'reality', or essence, be understood as bound up with 'actuality' as that term was defined above. The term 'mode of existence' is the suitable one to invoke here only once it is perceived that nothing static – as in, for instance, Spinoza's notion of *conatus* – inheres within it. Mode-of-existence (form) explodes ontology (that of Spinozistic 'modes' included) in every case where ontology privileges being over the forms it takes. Immediacy – the self-presence of being in, for instance, Heidegger (cf. Adorno 1973) – stands condemned. Through and through, form is dialectical. But denounced in the same breath as we declare the equivalence of form and mode of existence is every species of dialectics which, abstracting in the empiricist sense from practice, enunciates universal laws.

Abstraction *could not* exist concretely were the notion of 'form' just explicated to be dismissed. And 'form' *does* impinge on action, which makes Marx's declaration to the effect that alienated labour (activity) is the condition of private property (a social institution: passivity) all the more prescient for his later writings, and profound. Conversely, practice exists – in the above sense, it counts as ecstatic – through its forms; and, to be sure, most of these forms have been alienating. Practice during the entire course of recorded history (cf. the opening sentence of the *Communist Manifesto*) has subsisted solely in the form, or mode, of being denied. Discourse thematising concrete or practically-existing abstraction is accordingly discourse which introjects practice into reified categories. *Determinate* abstraction and *practically-oriented* abstraction go hand in hand. Hence whatever such abstraction does, theoretically, it is far from constituting theory as an 'independent' – a practice-independent – realm.

For the empiricists, however, an argument along these lines may not go sufficiently far. Theorisation in the mode just sketched is liable to appear 'formless'. It lacks specific focus (cf. Jessop's charge against Holloway,[20] and Althusser's polemics on 'essentialism', or in other words the reduction of species to genus: Althusser and Balibar

1970). How can a Marxism turning upon the notions of form and mode of existence and determinate abstraction address conjunctural questions? To enquiries of this kind it would be far too simplistic to say (what is indeed true) that 'conjuncture' belongs within a genus/ species discursive world. What is asked after is Marxism's purchase on the *real*, or actual, world and on this score empiricism has its own version of mode of existence to state. Emphatic traces of form as mode of existence can be discerned within the discourse of empiricist abstraction inasmuch as, in common with the nominalist tradition from which it springs, empiricism regards species as more real than genera and particulars as more real than both. Here, as in all nominalism, there is a serious move towards the rehabilitation of the particular which – however – does not go far enough. Particulars are brought into focus only so that they can *fall beneath* universals once more. It is as though someone were to say: 'individuality *exists already*, under capitalism', thus omitting (and thereby *sotto voce* writing an apologia for) the alienations which individuality suffers at capital's hands. *Form*-critique is accordingly the key to *soi disant* individualistic theories. In the same way, the emancipatory and nominalist current within empiricism can receive its due only once empiricist abstraction is replaced by abstraction of a determinate kind. The concrete (the conjuncture) subsists only as a contradictory impacting of abstractions, says the 1857 Introduction of Marx. In other words: once the linkage of abstraction to genus/species schemata is severed, the charge of 'formlessness' (e.g. Jessop 1991) is rendered impotent or, rather, reversed. For empiricism can address conjunctures only insofar as the latter 'fall beneath' something or other; only, that is to say, insofar as they can be assimilated upwards (the intriguing side of Althusser's anti-essentialist critique). Conjunctures construed *as a constellation of determinate abstractions* retain a sharper focus in the sense that the abstractions needful for the construal of the conjuncture are seen as inhering in (indeed as constituting) the conjuncture itself. The conjuncture is shown in sharp focus because it falls beneath, and is assimilated to, nothing. To this degree a helping hand is extended towards empiricism. But the hand is withdrawn at the same time as the implications of this helpfulness are unfolded: (a) there is in the end nothing within the idea of determinate abstraction to which empiricism might approximate and (b) once conjunctures are construed as constellations they cease, in the empiricist sense, to exist.

To render the above defence clear two definitions are required:

'determinate abstraction' and 'form'. Both definitions are implicit in what has been said earlier. Determinate abstraction is abstraction *in and through* which phenomena obtain, unlike empiricist abstraction which is *abstraction from* the phenomena concerned. Numerous empiricist abstractions indeed *are* determinate abstractions, but only in a fetishised or reified world: this is the import of Marx's critique of political economy. Existence-in-practice (however alienated the practice) is the criterion which determinate abstractions must meet.

Form, as above proposed, is to be understood in the sense of *mode of existence*. 'Form' presupposes 'determinate abstraction' and *vice versa*, inasmuch as the abstract can be the mode of existence of the concrete (and the other way round). Signalled, here, is the notion of an internally related 'field': anything can be the mode of existence of anything else (which is not of course to say that everything *is* the mode of existence of everything else, in fact). Moreover, just as the *terms* which are thus internally related are various – abstract, concrete, universal, particular, etc. – so too can be the *modes* in which they subsist. One limiting case is in this regard especially important: a term may exist (cf. Gunn 1987b) *in the mode of being denied*. That is, one term may exist in and through another *which contradicts it*. This, I take it, is the key to Marx's notion of commodity fetishism. When we learn that social relations which appear as 'material relations between persons and social relations between things' appear, thus, as 'what they are' (Marx 1976, p. 166) we are being informed of a circumstance that is unintelligible unless the notion of existence-in-the-mode-of-being-denied is taken on board.

This understanding of form as mode of existence is the central thematic of Hegel's *Science of Logic*. Space forbids an elucidation of such a contention; however, I suggest that it is in this sense that Hegel's insistence that his logic is also a metaphysics, or ontology (Hegel 1969, p. 63; *Encyclopaedia*, para. 24), is to be read. Thought-forms *qua* determinate abstractions are modes of existence. And Marx's declaration that 'in the *method* of treatment the fact that . . . I again glanced through Hegel's *Logic*' (Marx to Engels, 14 January 1858: Marx/Engels n.d., p. 121) is, surely, to be received in the same light.[21]

Empiricist abstraction's tendency is to favour a discourse of *external* relations: particulars stand one to another indifferently, and are united only through species (just as the relation between species obtains only through a genus which stands over and above). Determinate abstraction, by contrast, and as just indicated, points towards *internal* relatedness inasmuch as an internal relation obtains between A and B wherever B is A's mode of existence, or 'form'. If now we think of A as also (perhaps) B's mode of existence, and C as (perhaps) also A's mode of existence, and D as (perhaps) a mode of existence of B while also having its own mode of existence as A, and so forth, we discover a criss-crossing field of mediations which amounts to a *totality*: no term in the field stands as its own. This reinforces my earlier contention that 'totalising' historical materialism[22] stands in a closer relation to Marxism than does 'causalist' historical materialism (causalist theories always requiring, for their intelligibility, an external-relatedness of cause and effect). But equally it reinforces my contention that totalising historical materialism points *beyond* historical materialism, i.e. beyond 'sociology' construed as any general social theory whatever, since determinate abstraction explodes the notion of 'theory of' upon which empiricist abstraction rests.

This last point calls for further discussion. First of all, totalising theory *requires* the notion of determinate abstraction. *Minus* this notion, the conception of a 'mutual interaction' taking place between 'different moments', as is the case with 'every organic whole' (Marx 1973, p. 100) amounts to banality: everything somehow affects everything else (cf Williams 1973). To put bones into the flesh of totality we need to understand how terms can exist not just through one another but *in and* through – or, better, *as* – one another. We need to understand how terms can form and reform, or constitute and reconstitute, other terms: how one term's mode of existence can be another term, *without remainder*. This logically stronger conception redeems totality, and 'dialectics', from the vague notion of mere reciprocal interaction (an action which could be causalist or not as the case may be). *Not all interactions* count as totalising, on this logically stronger approach. Of course – in the light of the above argument it should go without saying – a totality can exist in the mode of being denied, for its part.

Secondly, the notion of 'theory of': the programme of empiricist abstraction goes something like this. Once we are in possession of the generic concept we can have a theory of species (for instance, once we know the genus we can begin to frame causal explanations premised on universal laws: lightning belongs under the genus 'electrical discharge' rather than 'God's anger', etc.); metatheory underpins the generic concepts in their turn. A proper fit, or mapping, as between concept and object is the result. We have seen already that this idea of theoretical consonance is what defines general theory. What has now to be made clear is the, so to say, dissonance of determinately abstract (totalising, 'dialectical') theory in this regard. Determinately abstract theory can be a 'theory of' nothing because it situates its own terms within the practical field that it reports.[23] It abnegates what Hegel terms the standpoint of 'consciousness' (see above). There is nothing standing over against determinately abstract theory for it to report. It throws the movement of its concepts into the crucible of its object (practice or 'actuality': see earlier) at every turn, thus undermining the separation of 'concept' and 'object' upon which ideas of consonance or mapping turn. General theory projects a theory (or metatheory) *of* its object first and a theory *in* its object only second. Determinately abstract theory reverses this priority, so to say setting *out* from inside. This being so, positivism is quite right, from its own point of view, in pouring ridicule on dialectics. Much more suspect are the dialectical theorists – as it were 'from Engels to Bhaskar' – who reckon that some rapprochement between different species of abstraction can be reached.

My argument has a final step. Does a shift from empiricist to determinate abstraction – from general theory to totalising theory, from historical materialism or sociology to Marxism – serve only to place theorising at the mercy of contingent practical winds? Once 'theory of' is dislodged from its pre-eminence, who can say on what shores we shall finish up?

In fact determinate abstraction, for better or for worse, carries with it a specificity of its own. That is, we can prescribe from it as a metatheory to the first-order theory it secretes. In fact a dilemma operates. *Either* we fall silent as to the first-order consequences of abandoning general theory (just as, and perhaps in part for the same reason that, Marx refuses to prescribe communism) *or* we concede – and this was always positivism's countercritique – that to condemn empiricism is to court irrationalism, whether *sotto voce* or in full voice. To escape this dilemma is to become unfaithful to ourselves, through prescription; to remain in it is to confess defeat. To resist both horns of the dilemma is to renew the theory/metatheory, and hence the theory/practice, split. Even supposing a reader to have found the argument so far compelling, this amounts only to a compulsion to cut the ground from his or her feet.

In fact *no* answer to objection along these lines is possible. Determinately abstract theory theorises *nothing*: it is – and this is our answer in effect – the only species of theorising which can. From 'nothing' to no-thing, i.e. to nothing which can ever be determined, conditioned or pinned down: instead of taking refuge in paradox (see earlier) this may seem like seeking refuge in a pun. No 'theory of' nothing is possible, self-evidently, as everyone from Parmenides and Plato onwards has declared for two thousand years. Nothingness has been confronted with the awful dilemma: either it really is something (in which case it contradicts itself) or it is indeed nothing (in which case, from Plato through Augustine, it is a privation of being, or of the good). So far as I am aware Hegel was the first to break with all such interdicts: man is 'evil' by nature, he said, in the course of discussing the Fall (*Encyclopaedia*, para. 24) and by this he meant not to that humankind is sinful within God's framework but that humanity *qua* negative escapes forever (or always-already) God's clutches. For it is only 'something' – some thing, in the mode of fetishism – that God can create. In prefering *no* answer to the above objections, therefore, we hardly concede the case. Nothingness *escapes* empircist abstraction, and our business is to trace the escape-route. Empiricism and theology construct fetishes (for instance 'society' as a reification), and acknowledging this simplifies matters wonderfully. The objection to tracing the escape-route ceases to be conceptual. It is just a matter of not telling our jailors, in advance, where we mean to go.

Determinate abstraction *prescribes* the terms of first-order theory only when looked at from the jailors' point of view. Determinate abstraction *is* first-order theory (while also being metatheoretical), and *vice versa*: here is the space of spontaneism.[24] The rigours of the first-order theory come into view, according to Marx, alongside communism, characterised neither in *The German Ideology* nor in *The Civil War in France* as a future estate of man but, in both, as the practically reflexive site of theorising: the real or actual movement of the working class. All discretely metatheoretical levels are rescinded. The prolepsis of the workers figures, methodologically (but even methodology stands detonated) as theorisation's common sense. ('Common', here, refers to theory's and metatheory's unity.) Far from amounting to irrationalism we discover here the most severe – because the most open-to-all-comers – test before which theorisation can ever be placed. Truth-claims can be raised only on the condition that *you can say (against them) what you like*.[25]

The nothing which practically reflexive or determinately abstractive theory theorises looks something like this: it consists in contradiction. From empiricist abstraction's point of view discourse which addresses itself to a reality (or actuality) viewed as contradictory can amount only to incoherence, or arbitrariness, because from a contradiction any proposition whatever can be deduced (Popper 1963, p. 139). And this is certainly right: programmes seeking to defuse 'dialectical' contradiction into some species of non-formal contradiction (e.g. Cornforth 1968; Gunn 1973) miss the point that if A *is* the mode of existence of B then A is A and not-A at the same time. (In Hegelian terminology: the 'motionless tautology of "I am I" ' – Hegel 1977, p. 105 – fails to report the self-movement of self-consciousness.) Contradictions are real, or rather actual. Empiricist abstraction, from Thales' contention that everything is *really* water onwards, has sought to defuse contradiction by assimilating it to the *difference* (the reciprocal *in*difference) of terms hanging together in some genus/species string. It *has* to do this since the premise of a 'theory of' is that everything be what it is and not another thing. (A theory *of . . .* what?) Determinate abstraction *has to* accept the actuality of contradiction since, as reported, if A can be the mode of existence of B then A can be not-A: there's no way out. Thus dialectical or totalising (Marxist) thought has to meet Popper's objection head-on. The *only* way in which it can meet it is to aver: from the indefinite range of what can be inferred from a contradiction *only some* inferences are to the point. The 'pointfulness' of the inference is to be decided by the subject-matter in hand (a concept/object unity once again).[26]

And so the defence against irrationalism comes down to a discussion of determinately abstract theory's subject-matter. By alluding to autonomist theory and to Marx's conception of communism we have already sketched a preconception of this: it is nothing other than the working class. But, as it happens, we can do better than to offer preconceptions. Basing ourselves on Hegel's and Marx's prolepses, we can say something about how contradiction appears. The following is what I have characterised as a wager, and no more.

All contradiction is complex, and triadic. For entirely non-formalistic reasons all contradictions go in threes, at least. For a contradiction can be contradicted only *by a contradiction*: were no thing (a contradiction) to be contradicted by something the outcome would be incoherence or stasis rather than the movement – the 'can be contradicted' or *modus vivendi* (Marx 1976, p. 198) – of the contradiction itself. In the same sense – perhaps in the same spirit – Spinoza declared that only freedom can delimit freedom. The same point obtains conversely. A contradiction identified as real, or actual, is one which 'has power to preserve itself in contradiction' (Hegel, *Encyclopaedia*, para. 382). 'The life of spirit is not the life that shrinks from death . . . but rather the life that endures it and maintains itself in it' (Hegel 1977, p. 19). In other words *contradiction-contradicted* is, at least potentially, the condition of contradiction itself. So far we have two out of our notionally three terms.

But even this two-fold version of contradiction can do logical – ontological and political – work. Contradiction (i) is what Hegel (1977, p. 101) terms 'this absolute unrest of pure self-movement' and Marx the real movement of the working class. That is to say, the set of a theorist who for present purposes may be termed Hegel/Marx is towards being which is self-determining. Marx's comments on 'species being' in the *1844 Manuscripts* say nothing else. Self-determination entails actually existing contradiction inasmuch as such being is what it is not (yet) and is (not yet) what it is (Kojève 1969, p.7; Bloch 1986, Intro.). Conversely, only in regard to the subject-matter of self-determining being does it makes sense to impute contradiction. Within the order of nature a stone is a stone is a stone (it is *what it is*) unless we kick it. Once we kick it we turn nature into a 'societal category' (Lukács 1971). But if the pre-kicked stone is already construed as contradictory, as in, say, Engels (cf. Gunn 1977),[27] then nothing but an 'animist projection' (Monod 1972) can be the result. For it is the existing-not-yet dimension of being which gives sense to the notion of real contradiction. Only beings with projects can so subsist. Nature being a societal category the human-kind/nature interaction counts as human for its part (Schmidt 1969). The political conclusions to be drawn here run as follows: neither the categories of a technological violation of nature nor of a non-violating stewardship of nature make any sense. We are out here on our own.

Humanist Marxism[28] stops with the level of contradiction (i): self-determination when focused on alone turns into self-gratification. Contradiction (ii) – the contradiction which contradicts contradiction *qua* self-determination – brings to light the force of what Marx calls the workers' 'self-emancipation' and also what he terms self-criticism, in the opening pages of the *Eighteenth Brumaire*. Throughout the history of political thought contradiction (ii) has generally been counterposed against contradiction (i): for instance in the distinction drawn between freedom and emancipation in Arendt (1973). The same distinction underlies Rousseau's famous 'forced to be free' paradox. The question underlying Arendt's and Rousseau's reports is that of how we can move from an unfree to a free state. If we move *freely* then we were not unfree to begin with, but if we move *unfreely* then freedom (at any rate in the sense of self-determination) can never be the result. Arendt's distinction and Rousseau's paradox are at any rate more rigorous than Hegel's conception (*Encyclopedea*) of the transition as one from potentiality into actuality: if, in effecting the transition, potential freedom plays a role then it is actual already; if it has no role to play then it is non-potential and non-actual at the same time. The trick has to be to see unfreedom as a mode of existence of freedom. Empiricist abstraction can contribute nothing to an understanding of how (uncontradicted) freedom might come about.

From the standpoint of determinate abstraction the matter can be viewed as follows. Contradiction (i) is the contradiction in which freedom, *qua* self-determination, consists. When this contradiction *stands contradicted* (contradiction (ii)) this amounts not to *un*freedom, stated literally, but to *unfree freedom*: to freedom-contradicted, or to freedom subsisting alienatedly, i.e. in the mode of being denied. (And only freedom *qua* self-determination, or as contradiction, can do this.) *For this reason* freedom has the 'power to preserve itself' (Hegel). Unfreedom subsists solely as the (self-contradictory) revolt of the oppressed. There is no question of climbing from unfreedom into freedom, as in Rousseau and Arendt, because the former is non-existent: communism obtains *already* as the real movement of the class. Potentiality/actuality distinctions are overturned along with stages of history and empiricist abstraction. Stated otherwise: contradictions (i) and (ii) *are the same* while maintaining their difference intact. If they were completely the same there would be no distinction between freedom and unfreedom whereas if they were completely different the ascent from unfreedom to freedom would be impossible. Each is *and is not* each other. Each is the other, but in the mode of being denied.

So far, we have remained on more or less humanistic ground because only two moments of contradiction have come before us. If we halted at this point the consequence, rightly identified by Althusser, would be an essentialism of contradiction (i). At most we would have a *vulgar* spontaneism – 'Just get on to the streets!' – no less fetishising contradiction (ii). Althusser's signal and crucial shift is to what I shall call contradiction (iii): the contradiction *in which the contradiction which contradicts contradiction consists.* That is, contradiction (ii) relates two terms each of which is contradictory for its part:[29]

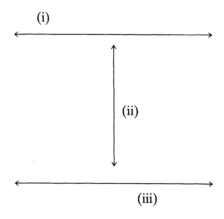

(i)

(ii)

(iii)

Were any one of these terms to be something rather than no thing (rather than a contradiction) the space of the 'field' or totality – indicated in the above diagram by the outward-pointing arrows – would collapse. For a contradiction can 'preserve itself' only when contradicted by a contradiction. In other words it is contradiction by, precisely, a contradiction which allows a contradiction not sheerly to detonate but to 'move' (Hegel 1977, p. 101; Negri 1984; Marx 1976, p. 198). This is not to say that a contradiction contradicted by *something* simply dissolves. But for it not to dissolve its relation to the non-contradictory 'something' has to be understood as contradictory on its own account (for instance the humankind/nature relation has to be seen as a *social* relation). And that means that the notion of contradiction *per se* has to be addressed first, as above. Unless these three moments of contradiction are thematised contradiction vanishes: for each implies the other. Each *is and is not* the other. Were contradiction to consist in (i) alone it would *be* itself, purely, which is to say that it would not be (exactly positivism's charge). For it to be *and* not be – for it to subsist in its own mode, i.e. as contradiction – (ii) has to come into play. However a contradiction contradicted by nothing would remain something, i.e. a non-entity; and so contradiction (ii) must carry in its wake contradiction (iii). Contradiction (iii) counts as *something* from the standpoint of contradiction (i): for instance the whole weight of social structures oppresses individuals struggling to be self-determining, or free.

Durkheim, thus, focuses on contradictions (i) and (iii) alone, in abstraction from contradiction (ii), when in the *Rules of Sociological Method* he reports social facts as coercive. Durkheim gives the somethingness of (iii) its full weight. However if (i) and (iii) are related contradictorily then (iii) is shot through with contradiction for its part. Social worlds comprise not just a 'dead essence', reports Hegel (1977, p. 264); each of them is *'actual* and *alive'*

Althusser of course picks up on just this liveliness. Hence his notion of a process without a subject which, abstracting (iii) from (i) and (ii), he goes so far as to attribute to Hegel (Althusser 1976). Althusser thus confirms the alibi of a *sociology*: humanists (first contradiction) are left floundering and autonomists (second contradiction) remain merely on the threshold of social 'theory of'. For undoubtedly contradiction (iii) taken on its own focuses exactly on the 'society effect' (Althusser 1970). Social structures flourish, but whether the flush on their cheeks is that of health or fever, or of life or death, remains undecidable once contradiction (iii) is taken as the sole theme. Unsurprisingly 'alienation' is a category which Althusser – I cite him only because he is the most rigorous of sociological Marxists – hurls away. Unsurprisingly the humanists (McLellan *et al.* in the 1970s) champion alienation and simply shout Althusser down. The whole debate was, of course, useless: contradictions (i) and (iii) can be brought into relation *qua contradictions* only through the mediation of contradiction (ii).

At this point a caveat has to be entered. It was Kant in the *Critique of Judgement* who first declared for the logical (and *sotto voce* the ontological) primacy of the number three. Behind this declaration lie, of course, Augustine and Joachim of Fiore and so on. Hegel in his introductory material to the *Science of Logic* urges – a small voice in a wilderness – that 'three' should not be taken in a formalistic sense. The uprising of Engels changed everything. Not only did Engels, in the *Anti-Duhring* and the *Dialectics of Nature*, propound the teleological notion of thesis/antithesis/synthesis: by an amazing coincidence he discovered three (and only three) dialectical laws. Triadism, subsequently, never managed to recover its dignity from such a bathetic celebration. Since I have declared contradiction within its own internal structure to be triadic it therefore behoves me to say: no one of the three contradictions *constituting* contradiction has primacy. No synthesis is to be discovered within what I have termed contradiction's 'field'. The numbers (i), (ii) and (iii) are purely nominal. The same argument could have been constructed no matter what entry-point we assumed. Teleology vanishes the moment contradiction, thus understood, appears.

And subsequent to the caveat, a claim: the above exposition of contradiction remains unintelligible where only empiricist abstraction, or general theory, rules. It remains unintelligible because to make the matter clear we have to say that each contradiction is *the mode of existence* of each other. Each is the form (each re-forms) each other. None, separately, can exist without such a re-formulation: in theory and practice as well. And returning to an earlier theme in the chapter we have to say, also: there is nowhere outwith such a field of contradiction where (practical reflexivity counting as operational) any theorist of society might stand. Determinate abstraction mediates practical reflexivity into contradiction and *vice versa*: particulars and universals dance, whether or not to the same tune. 'Form' has to be understood as mode of existence rather than as species (with all the unpredictability and anarchism that modes of existence, construed to be sure proleptically but without teleology, may take). Empiricist, or general-theory, abstraction takes 'form' as something fixed, or at any rate fixable; determinate abstraction takes it to be movement and as nothing (no thing) else. A secondary claim should be, here, entered. The unfixity of form signals its openness to a future.[30] This openness is *intrinsic* to contradiction. Therefore contradiction (ii) takes priority over contradictions (i) – humanism – and (iii) – structuralism – alike. Ultimately, this is why the autonomists of the 1970s were on the right track.

I close with a number of theses, each entailed by the argument I have sketched. Each one of these theses is true:

- Marxism is the critique of general social theory rather than its confirmation; therefore Marxist sociology, along with historical materialism, should be condemned if not ignored.
- In more detail, totalising Marxism is more rigorous than the causalist version: but no comfort for either humanist or structuralist Marxism is preferred by this thought.
- Whereas general theory stands back from its object and reflects upon it, Marxist theory situates itself within its object (practical reflexivity) and construes itself as constituted through its object (determinate abstraction).
- That is to say Marxist theory *has no* object. For a theory to have an object is for it to project an adequate relation between its object and its concepts (cf. Hegel 1977, Intro., on being-for-

consciousness and being-in-itself), a relation requiring to be metatheoretically underpinned (cf. Hegel 1977 on the notion of 'criterion'). *If one likes* one can say that Marxism has the movement of contradiction as its object; but such a formulation is more misleading than helpful since the peculiarity of contradiction (in its threefold guise) *is that it moves*: it is precisely that which, in contrast to 'objects' – in all the received ontological and epistemological meanings of that term – can never be pinned down. All 'mappings', including the mapping of concepts on to objects, require at least a relative stasis on the part of that which is mapped. Marxism demythologises stasis. A static society is for Marxism a special instance, i.e. an instance of alienation or fetishism; such a society is to be construed as contradiction subsisting in the mode of being denied.

- Stated schematically: structures are modes of existence ('forms') of struggle. To say this is not to fall into an *essentialism* of struggle, however, because contradiction (iii) is just as real as contradictions (i) and (ii). Nor, conversely, is it to endorse an eclecticism whereby (cf. Jessop 1988) structure and struggle are reciprocally conditioning; for contradiction (iii) *is* contradictions (i) and (ii). It is, and is not. Between the three contradictions precisely – as it must do – a contradictory relation obtains.

- It follows from this that our three contradictions are far from amounting to three 'levels' of abstraction, in the empiricist sense. Each is a form of each other but none is a species of each other. Each is the other's mode of existence. Inasmuch as contradiction *is threefold* it follows that contradiction *becomes invisible* once empiricist abstraction is brought into play.

- All of the above follows from the enunciation of practical reflexivity, which allows a sort of common sense of theory and metatheory to break out. The notion of a theory *having an object* depends on the theory/metatheory split. Concept and object spring apart (bringing the programme of 'mapping' into play) in the same movement as do theory and metatheory. A unity of theory and metatheory situates the concept in the object and *vice versa*.[31] The standpoint which Hegel termed that of 'consciousness' is surpassed. 'Objects' are detonated. Doubtless theory belonging within what I have called empiricism and relying on the theory/metatheory severance sometimes alludes to what resembles practical reflexivity (e.g. Bhaskar *et al.* 1988; cf. Gunn 1989a, p. 94–5). Such an allusion achieves nothing, however. It ignores all the

innovative opportunities and consequences for theorising along a
first-order axis which the present paper has sought to clarify.
Practical reflexivity places *what counts as theory* up for grabs.
- The distinction between *social theory* and a *theory of society* (a
'sociology') is all important: the latter reifies society as something
standing out in front of the theorist whereas the former construes
the constitution of categories as occurring behind the theorist's
back. The latter depends on a subject–object split (the standpoint
of 'consciousness') whereas the former invokes reflexivity, viz.
reflexivity of a practical kind. From the former's point of view the
latter's objectivity appears fetishised (cf. Backhaus, Volume One)
whereas from the former's point of view the latter appears
metaphysical – but only because empiricist abstraction is already
presupposed *a priori*.

*A*social critiques of theories of society are two-a-penny. Society
does not exist from, say, a methodological individualist point of
view. But the individualism of, for instance, an Elster (1985) or a
Roemer (1986) has to bend over backwards to discern even the
faintest traces of the social and intersubjective constitution of the
individual upon which the discourses of a Hegel or a Marx turn. (For
of course 'the individual' is a determinate abstraction once more.)
Social critiques, on the other hand, have been all too sociological:
Althusser and his structuralist epigones translate social theory into
(fetishised) *theory of*. Between social theory (Marx) and sociological
theory (the epiogones: not just the structuralist ones, but the
individualists too in the sense that even the latter announce a
rapprochement to 'society' in terms of equilibria or a law of unin-
tended consequences as the case may be) a wedge therefore has to
be driven. The illusion according to which social theory must
automatically be sociological theory has to be brought into question,
relfexively. The illusion which declares that non-sociological theory
has to be methodologically individualist requires challenge, no less.
The wedge has to be driven in at the appropriate angle: a highly
precise one. The geometry of such a theoretical and practical
intervention is what the present chapter has aimed to make clear.

Notes

1. Neither Marx nor Engels employed the term 'historical materialism'.
   Engels (Marx 1971, p. 220) favoured the locution 'materialist conception

of history'. No equivalent locution is to be found in Marx. Despite this, I think that *historical materialism* is a reasonably accurate designation of the version of general social theory into which Marx (occasionally) and Engels (always) lapsed.

2. Much cited as historical materialism's *locus classicus*. The citation is the more remarkable because it comes from all sides. According to Lenin we have here 'an integral formulation of the fundamental principles of materialism as applied to human society and its history' (n.d, 21, p. 55). (N.B.: the idea of 'application'.) According to Eduard Bernstein's commentary (1909, p. 3) – although it turns out that Bernstein bases himself more closely on Engels's letters to Bloch, Schmidt *et al.* of the 1890s – 'No important thought concerning the Marxist philosophy of history is wanting here'. Plamenatz (1970, p. 18, 19) sees the Preface as Marxism's 'classic formulation'. According to Michael Evans (1975, p. 61) the Preface is the 'classic account' of Marx's 'general conclusions'. Cohen (1978) declares for the same view.

3. See for example Althusser 1970, p. 59 who declares that sciences only once they are 'truly constituted and developed' can supply their truth-criteria. But what counts as a true constitution? Does first-order theory itself determine it? Althusser proclaims Spinoza's view according to which truth *per se* is the indication (*index*) of truth and falsity. Can then scientology, for example, or UFOlogy testify in their own case?

4. From Althusser's *For Marx* through his *Lenin and Philosophy* into his *Essays in Self-Criticism* we can read the self-humbling of philosophy as metatheory. It started off as the 'Theory of theoretical practice' and finished up as 'class struggle in the realm of theory'. And yet, despite this humbling, philosophy remains, in Althusser's view, the decisive instance within 'dialectical materialism'. The upshot is a historicism and relativism (see also previous note) which is the other side of the coin of the structuralist 'science of history' which Althusser sets out to defend.

5. Transcendental arguments become viciously circular when they attempt to *ground* categories: if X presupposes Y this in no way demonstrates either Y's validity or (as an inference from Y) the validity of X. Hence relativism and historicism again. Transcendental forms of argument have a respectable role as *ad hominem* arguments but Bhaskar, although presenting himself as an underlabourer (cf. Bhaskar 1989), apparently wants them to achieve more.

6. Application, however 'creative', inheres within a genus/species framework. And in case it should seem that, in urging a genus/species framework to be un-Marxist, we are contending against a straw man it should be noted: an entire school of Marxist orthodoxy (dialectical materialism or DIAMAT) made this framework its own. Dialectical materialism was seen as a species of the genus 'materialism', and historical materialism was understood as a species of materialism of a

dialectical kind. Communist Party education classes, and the readers for them, were organised in just this Aristotelian way. The first moves against this genus/species scheme were made by Lukács's *History and Class Consciousness* and Gramsci's *Prison Notebooks*, which privileged historical materialism over dialectical materialism and history over materialism (e.g. Gramsci 1971, p. 454–6).

7. Acton 1955; Plamenatz 1970, ch. 5, section 2; Cohen 1978, ch. 5; Lukes 1984; and so forth.

8. Poulantzas 1978; Panzieri 1976.

9. According to Marx the 'forces of production' include natural science. Additionally, they include 'general social knowledge' and 'the human being himself' (Marx 1973, p. 706, 422). Labour is itself 'the greatest productive power' (ibid., p. 711). Also included are the powers of cooperation and of intercourse in *Capital* and *The German Ideology*. From a general-theory standpoint the difficulty is less to know what does than what does not count as a productive force. And from the same standpoint just the same difficulty applies to productive relations. In *Wage Labour and Capital* (Marx 1967, p. 28) we learn that *all* social relations are relations of production – that is, not just 'basic' or 'economic' relations – should we construe them in that way. Forces and relations of production therefore seem to interweave. Each can count as the other. Should we see this as a strength or a weakness in Marx's thought?

10. Either/or formulations are always problematic, and so a word of caution should be entered. The either/or contrast refers only to theory's starting point, the reduction of structure to struggle being just as obnoxious as its converse. The dialectical question here concerns the categories 'form' and 'content', which for reasons of space this chapter is able to comment upon only briefly. The only question can be that of mutually exclusive problematics. A dialectics of subjectivity is able to encompass objectivity as reification, whereas a dialectics of objectivity expels subjectivity from its consideration and can re-engage with it only via decisionism or historicism. Hence there is no question of a reductionism here.

11. See the definition of 'structure-in-dominance' in the glossaries of all of Althusser's books. Whichever moment or practice or instance is dominant within some epoch figures as economically determined – in the last instance. Ultimately Althusser appears solely as a commentator on Engels's late letters. Do non-economic factors appear as independent variables (however slight) or just as local disturbances? Can they make any real difference to history's vector (Engels's metaphor)? On this score neither Engels nor Althusser teach us anything approaching a clear view.

12. The text which Marx in 1859 declared to be left to the gnawing criticism of the mice was of course *The German Ideology*. The 1859 Preface

presents an intellectual autobiography ('A few brief remarks concerning the course of my study': Marx 1971, p. 19) culminating in a report of the 'guiding principle (*Leitfaden*)' at which Marx arrived 'in Brussels', the place of *The German Ideology's* drafting. The famous 'general conclusion' at which he arrived, and which the 1859 Preface summarises, is accordingly that reported in *The German Ideology* (esp. Part One). Two comments are in order here. First: the summary contained in the 1859 Preface *has no scientific status*, not just because Marx who had re-read Hegel's *Logic* in 1857–8 may well have shared Hegel's view of prefaces – a view apparently endorsed by Engels when in his correspondence of the 1890s he says that only a *Hauptprinzip* (cf. the introductory material in Hegel's *Science of Logic*) can be recovered from Marx's 1959 views – but also because it is only at the *end* of the 1859 Preface that, according to Marx, 'science (*Wissenschaft*)' erupts. Secondly, and more importantly, the discourse of *The German Ideology* is falsified, consciously or unconsciously, by the Marx of 1859. A reading of the section of *The German Ideology* Part One entitled 'The Premises of Historical Materialism' – an editor's subtitle if ever there was one – reveals an *ambiguity* in Marx's 1845–6 approach as between totalising and causalist views. The premises concerned are reported as those of the production of the means of subsistence, of 'new needs', of sexually reproductive structures, of 'a certain mode of co-operation', of language (and hence consciousness): all in an ascending order (Marx/Engels 1975, Vol. 5 p. 41–44). This notion of ascendancy is *causalist* on a grub-before-ethics basis, just as was Engels's speech at Marx's graveside, where, in Social Darwinist fashion, the necessary and sufficient (causalist) conditions of social development became so fatefully confused. But Marx adds (Marx/Engels 1975, p 43) that the 'aspects of social activity' he has mentioned 'are not to be taken as . . . different stages' but rather as distinguishable 'moments' of social existence *per se*. Here the perspective ceases to be causalist (grub conditions ethics) and becomes *totalising*; because once the production of the means of subsistence or of new needs becomes *social* (re)production, mediated for instance through consciousness, or self-reflection, nothing can ever be the same again. Precisely the 'ordering' of the alleged premises is problematised. The last might just as well come first. Social (reflective, antagonistic) production fails to equate with the unreflective production of bees, beavers or ants. In *The German Ideology*, this ambiguity between causalist and totalising views remains fertile, allowing Marx to pitch against each other the 'mechanism' of the eighteenth-century materialists and the 'teleology' of Hegel and Kant. In the 1859 rendition of *The German Ideology* insight, only causalism is remembered and totalisation drops out of view. Marx himself may have had good (*Realpolitiker*) reasons for such a misremembering (cf. Prinz 1969) but his reasons can hardly be our own.

13. 'In the name of action . . .': here again it is indispensable that we become cautious, because action construed *sui generis* is so easily an empty form. *Form*-analysis is the key. Only in the name of action, I want to say (cf. Backhaus's critique of 'objctivity' *a la* Adorno: Volume One), can form-analysis break out. Even classical bourgeois political economy 'has never once asked the question of why this content has assumed that particular form' (Marx 1976, p. 174). My point relates to the categorical considerations which allow this difficulty to be posed.

14. In German the pun as between 'actuality (*Wirklichkeit*)' and 'activity' works just as it does in English (*vide* Hegel 1969, p. 546; Moltmann 1985, p. 313). Actuality connotes not so much reality (and certainly not a static 'given' reality) as practice. Marx's 'actual, active (*wirkliche werkenden*)' labourers in *The German Ideology* register the same point: actuality and activity are the same thing.

15. I mean it in Pascal's sense. The kitsch version of Pascal is to the effect that you might as well reckon yourself to be immortal because, that way round, you have nothing to lose. The less kitsch version of Pascal reports that you have to bet your reason against your salvation. The secular version of Pascal avers (but it's only a wager once again) that reason and salvation – philosophy and the proletariat – can be understood from the same set of cards.

16. Cf. Kant's *Critique of Practical Reason – passim*, or for instance, Thomas Reid's *Essays on the Active Powers of the Human Mind*. Always, free-will discourse has linked hands with causalism. Hermeneutical sociology (e.g. Weber) does the same thing inasmuch as it awaits a causal explanation for the interpretations preferred. Weber's *Economy and Society* sociologises action (i.e. it brings action under the sign of empiricist abstraction) in the same movement as it defines *social* action as action which takes account of others' views. Giddens's conception of 'human action' as action which manages to 'construct the social world' while also being 'conditioned or constrained by' the 'very world' of its creation (Giddens 1981, p. 54) inheres in the same tradition. So too does Bhaskar (1989, ch 5). The ancient dispute concerning voluntarist *versus* determinist sociologies (together with their putative interrelations) becomes needless, I want to say, once the terms of discourse are shifted on to a non-empiricist terrain.

17. To declare against genus/species frameworks in regard to the idea of a historical sequence of modes of production (or of social formations, in the more sophisticated structuralist version) may seem too easy a point to score. For how then are pre-capitalist modes to be theorised? The implication of my argument is to the effect that they can be theorised only from within an approach which, *qua* practically reflexive, construes itself as sited on capitalist (i.e. socially contradictory) soil. For it is only then that ethnological abstractions – religion, the family, magic, econ-

omics and production itself – can be problematised. 'Primitive religion' or indeed 'feudal religion', for instance, remain senseless abstractions. For an illustration of how deep this problematising can go into the apparently socially neutral category of labour (for surely all societies must 'work' in order to produce their means of subsistence?) cf. Baudrillard 1975. In other words, the still-empirical category of 'production-in-general' employed in Marx's 1857 Introduction is *not* the key here. Marx in the same text goes so far as to say that 'there is . . . no general production' (Marx 1973, p. 86), as though pulling his empiricist *alter ego* – the author of the 1859 Preface – up short. On the origins of the 'modes of production' debate see Dobb 1946; Hilton 1978; Hobsbawm in Aston 1965.

18. In the 1872 version of the rules for the first Communist International.
19. Not even the notion of 'intermediate' levels of abstraction, as advertised by the Regulation Approach and Fordism/post-Fordism, can rescue it. Intermediacy inheres totally in – indeed it is the very definition of – abstraction which has taken a genus/species and thereby an empiricist form.
20. Jessop 1991 declares against the 'formless' character of class struggle as Holloway, for instance, reports it. Jessop for his own part announces a programme of 'fleshing out' the 'modalities' of class struggle: 'we will be misled . . . if we focus on class struggle without regard to its specific forms'. Some struggles, furthermore, can 'acquire class-relevance' and others cannot. A genus/species scheme, thus, is deployed by Jessop both *amongst* class struggle's modalities and *between* the modalities of specifically class struggles and struggles of other kinds. 'Form' is understood solely in terms of 'species' and not at all in terms of 'mode of existence' (see below). The notion of the *re*-forming of precapitalist struggles – over the family, for instance (cf Gunn 1987) – in and through the capital/labour class relation thus drops out of sight. Modalities are just species falling, statically, beneath a genus of some kind. The fundamentally *sociological* aspect of Jessop's discourse in the piece here under discussion emerges, likewise, in his comments on value. He refers to 'the dominance of the value form (or, better, since it has several different moments, meta-form)': the 'different moments' are for instance those of the 'commodity, price, and money forms'. In other words value is for Jessop a 'meta-form' solely because it is a genus containing numerous species – or 'forms' *in the sense of* species – within it. The horizontal extension of this reflection from species to genus which (counting as a renewed species) reflects into a further genus . . . requires, of course, some metatheoretical vertical ascent to sort matters out. Hence perhaps the eagerness with which Jessop (1988/9) embraces the Critical Realism of Bhaskar. Sociology (first-order discourse severed from metatheory) and philosophy (higher-order discourse severed from the theory it seeks to underlabour for, or command) are, I propose, two sides – each

requiring the other – of the same coin: either vicious circularity or infinite regress. And this coin has no currency of a Marxist kind.

21. The categories of universal/particular, form/content, etc. upon which the present chapter reflects are all Hegel's. Were my paper to be more than a sketch, either a rewriting or a critique of the *Science of Logic* – which throughout dwells upon the notion of determinate abstraction – would be the outcome. See e.g. Marcuse 1987. On one Hegelian category, that of quantity/quality, I permit myself further comment. Engelsian Marxism is famous for having declared that quantitative change (at some point) passes over into qualitative change and *vice versa* (e.g. Engels 1964, p. 63). The parallel passages in Hegel's *Science of Logic* are much celebrated, and commented upon in Lenin's 'Philosophical Notebooks'. But then there comes in Hegel (1969, p. 375ff.) a passage where quantity is discussed not just as passing over into quality but *as* quality: on this Lenin finds it possible to remark only 'Transition . . . expounded very obscurely' (Lenin n.d., xxxviii, p. 125). Unfortunately for Engels and Lenin this 'transition' *is a crucial* passage in Hegel upon which Marx's *Capital* relies. For there, quantitative categories – value, surplus-value, etc. – are construed *as qualitative*, i.e. as issues in class struggle (cf. Cleaver 1979). Even the notion of just such an inverse ratio as that obtaining between necessary labour and surplus labour is selected by Hegel as his example showing how and why quantitative distinctions require construal as qualitative as well (1969, p. 378): here *if anywhere* – cf. MacGregor 1984 – the notion of surplus-value unfolds within Hegel's thought. Engels and Lenin failed to read what lay before their eyes. Quantity is for Marx and Hegel quality's mode of existence, and *vice versa*. Quantity transforming *into* quality and *vice versa* remains an all too external relation: Marxist sociology is its result, also bourgeois-Marxist economics (cf. Backhaus, Volume One. Quantity *as* quality (and *vice versa*) announces the critique of sociology – of general social theory – inasmuch as the externality of the relation between 'theory' and 'of' in the expression 'theory of' is problematised. Lenin's 'very obscurely' announces the defaulting of more than DIAMAT's putative Marxist thought.

22. See for example Lukács 1971, 5th essay; Marcuse 1941, p. 316–9; Sartre n.d., p, 34. Within this totalising discourse capitalism appears as the first *sociological* society in virtue of the fetishised character of its objectivity. (cf. Backhaus, Volume One, and Horkheimer 1972, p. 229 on 'mechanism' and machines). In other words, Marx at his most general (as in the 1859 Preface) is Marx at his most particular, or at least specific, when read in the most charitable way. *Only* capitalist society is historical materialist society, with all the determinism that entails. A *particular* focus thus underlies the most generic of Marxist abstractions. A historical materialism (a sociology) of communism, for instance, would be a contradiction in terms.

23. The term 'field' is, here, to be stressed. Outside of the field of contradictions which I discuss later on there lies nothing, or rather the dualism of 'inside' and 'outside' is overturned. *There is nowhere outwith a terrain of (practical) contradiction* for theory to stand: cf. Kojève 1972 who reports the same point *vis-à-vis* 'totality' in that term's Hegelian sense.

24. Spontaneism in the 'classic' sense of a movement of contradiction rather than in the 'vulgar' sense of sheer immediacy: 'On the one hand we have the day-to-day struggle; on the other, the social revolution. Such are the terms of the dialectical contradiction through which the socialist movement makes its way' (Luxemburg 1970, p. 128–9).

25. For a defence of this severely informal notion of rigour see Gunn 1989b. Gunn, in this piece, still relies far too much on the empiricist notion of a one-to-one mapping of concepts on to objects: 'theory of'. His more logically powerful insight is to the effect that a truth which hits its mark in a 'correspondence' sense is like a golf ball which scores a hole in one: neither the theorist who raises truth claims nor the golfer who raises his club need be able to say how their respective successes are achieved. The consensus version of truth which Gunn sets out to defend involves a German-philosophic principle of *reflexivity*: truth counts as such if and only if it can be defended, before all comers. Coincidental accuracy fails to qualify as truth, in this way. The deficiency of correspondence theories of truth is that they admit just such coincidences into their canon, and so open themselves to the charges of historicism and relativism raised above. Truth *involves* reflexivity. Truth *minus* reflexivity amounts to guessing. Reflexivity opens the path to determinate abstraction, and determinate abstraction the path to society of mutual recognition wherein (*Communist Manifesto*) the freedom of each is the condition, and the result, of the freedom of all. Gunn raises well-intentioned points, but these need to be bathed in the fire which determinate abstraction represents. Only if the freedom of each is the condition of the freedom of all can truth-claims, within the ambit of a (practical) reflexivity which requires some notion of their audience, be redeemed.

26. 'Pointfulness', here, requires a further dialectical twist. 'Points' appear epistemologically as all-too-well – as all-too-fetishistically and reifiedly – established. Across an epistemological world construed under the sign of such a fixity determinate abstraction drives like one more hurricane. To appreciate the point of the pointfulness which the sentence here noted invokes we have to be able to see that *if abstractions can be real* (*qua* determinate) then *societies can be true or false*, for their part. False societal abstractions can require true determinately abstract theory, if indeed their truth – the truth of their falsity – is to be reported well. 'In a world which *really is topsy-turvy*, the true is a moment of the false' (Debord 1987). The subject-matter-in-hand is one preformed in such a

way as to drop through the fingers which want to hold it, in any sociological sense. Cf. Horkheimer 1972 and Adorno 1973.

27. Gunn 1977 makes the awful mistake of construing contradictions as merely conceptual (= nominalist, = empiricist). This deficiency being subtracted, however, the rest of his argument more or less stands.

28. Here I refer not merely to McLellan, who within the exigencies of class struggle can appear overly anodyne, but to, e.g., the not-yet-existence of Bloch. Bloch has the sheer human decency to imply that our future might be uncoloured by our past sins: immediacy, therefore; vulgar spontaneism despite Bloch's own Leninist politics. Contradiction (i), on its own, inheres in empiricist abstraction: people *in general*, whether they be apocalypticists or utopians (cf. Gunn 1985), are in the last instance nice. Cf. Holloway 1988 for the same Bloch-inspired optimistic view. The mangles of contradictions (ii) and (iii) still lie ahead, however, and so read on.

29. Cf. Gunn 1988. The diagram which follows is nothing but a literal rendition of the following two sentences in Hegel: 'Spirit is, in its simple truth, consciousness, and forces its moments apart. *Action* divides it into substance, and consciousness of the substance; and divides the substance as well as consciousness' (Hegel 1977, p. 266). *Everything* in Hegel's discussion of the alienations of spirit (1977, ch. VI) depends on this point.

30. For the sake of clarity three senses of 'form' should be distinguished: (i) form can refer to conceptual status (as in 'the form of Marx's 1859 Preface is at least as problematic as its contents'); (ii) form can refer simply to species (as in Jessop 1988/9); and (iii) form can refer to mode of existence, the sense of 'form' which must needs be rediscovered should anything approaching radicalism wish to break out. It goes without saying that 'mode of existence', in its turn, should not be construed statically. Existence = ek-sistence = ecstasis or ecstacy. We live ecstatically, ahead of ourselves, into and through our forms.

31. This point is developed eloquently in Hegel 1977, Intro., where it is declared that 'being-in-itself' has always to be a 'being-in-itself-for-consciousness'. The planes of theory and metatheory according to Hegel *intersect*, and the Kantian view (for example: cf. Bhaskar 1989) of cognition as an 'instrument' falls over its own feet.

References

Acton, H. B. (1955) *The Illusion of the Epoch* (London).
Adorno, T. W. (1973) *Negative Dialectics* (London).
Althusser, L. (1969) *For Marx* (London).

Althusser, L. (1970): his contribution to Althusser, L. and Balibar, E. *Reading Capital* (London).

Althusser, L. (1971) *Lenin and Philosophy and Other Essays* (London).

Althusser, L. (1976) *Essays in Self-Criticism* (London).

Arendt, H. (1973) *On Revolution* (Harmondsworth).

Aston, T. (1965) *Crisis in Europe 1560–1660* (London).

Bahro, R. (1979) *The Alternative In Eastern Europe* (London).

Bataille, G. (1985) *Visions of Excess* (Manchester).

Baudrillard, J. (1975) *The Mirror of Production* (St Louis, Miss.).

Bellamy, R. (1987) *Modern Italian Social Theory* (Cambridge).

Bernstein, E. (1909) *Evolutionary Socialism* (London).

Bhaskar, R. (1989) *Reclaiming Reality* (London).

Bhaskar, R. *et al.* (1988) 'Philosophical Underlabouring', *Interlink*, no. 8.

Bloch, E. (1986) *The Principle of Hope* (Oxford).

Bonefeld, W. (1987) 'Reformulation of State Theory', *Capital and Class*, no. 33.

Cleaver, H. (1979) *Reading Capital Politically* (Brighton)

Cohen, G.A. (1978) *Karl Marx's Theory of History: A Defence* (Oxford).

Comninel, G.C. (1987) *Rethinking the French Revolution* (London).

Cornford, F.M. (1939) *Plato and Parmenides* (London).

Cornforth, M. (1968) *The Open Philosophy and the Open Society* (London).

Debord, G. (1987) *Society of the Spectacle* (Rebel Press, n.p.).

Dobb, M. (1946) *Studies in the Development of Capitalism* (London).

Elster, J. (1985) *Making Sense of Marx* (Cambridge).

Engels, F. (1964) *Dialectics of Nature* (Moscow).

Evans, M. (1975) *Karl Marx* (London).

Geras, N. (1972) 'Marx and the Critique of Political Economy', in Blackburn (ed) *Ideology in Social Science* (London).

Giddens, A. (1981) *Historical Materialism: A Contemporary Critique* (London).

Gramsci, A. (1971) *Selections from the Prison Notebooks* (London).

Gunn, R. (1973) 'Marxism and Dialectical Method', *Marxism Today* vol. 17 no. 7.

Gunn, R. (1977) 'Is Nature Dialectical?', *Marxism Today*, vol. 21, no. 2.

Gunn, R. (1985) 'Notes on Utopia and Apocalypse', *Edinburgh Review* no. 71.

Gunn, R. (1987a) 'Practical Reflexivity in Marx', *Common Sense* no. 1.

Gunn, R. 1987b) 'Marxism and Mediation', *Common Sense* no. 3.

Gunn, R. (1988) '"Recognition" in Hegel's *Phenomenology of Spirit*', *Common Sense* no. 4.

Gunn, R. (1989a) 'Marxism and Philosophy', *Capital & Class* no. 37.

Gunn, R. (1989b) 'In Defence of a Consensus Theory of Truth', *Common Sense*, no. 7.

Hegel, G.W. F. (1892/1971) cited as *'Encyclopaedia'*: *The Logic of Hegel* and *Philosophy of Mind* (Oxford).

Hegel, G.W.F. (1969) *The Science of Logic* (London).

Hegel, G.W.F. (1977) *The Phenomenology of Spirit* (Oxford).

Hegel, G.W.F. (1986) *The Jena System 1804–5: Logic and Metaphysics* (Kingston and Montreal).

Hilton, R. (ed.) (1978) *The Transition from Feudalism to Capitalism* (London).

Hobsbawm, E.J. (1964) Introduction to Marx *Pre-capitalist Economic Formations* (London).

Holloway, J. (1988) 'An Introduction to *Capital*', *Common Sense*, no. 5.

Holloway, J. (1989) 'Marxism: A Theoretical and Political Programme', *Common Sense*, no. 7.

Horkheimer, M. (1972) *Critical Theory: Selected Essays* (New York).

Jessop, B. (1988) 'Regulation Theory, Postfordism and The State', *Capital & Class*, no. 34.

Jessop, B. (1988/9) 'Regulation Theories in Retrospect and Prospect', *Staatsaufgaben*, University of Bielefeld Zentrum für interdisziplinäre Forschung (Bielefeld).

Jessop, B. (1991) 'Polar Bears and Class Struggle', in Bonefeld and Holloway (eds) *Post-Fordism and Social Form* (London).

Kojève, A. (1969) *Introduction to the Reading of Hegel* (New York).

Kojève, A. (1972) 'The Idea of Death in the Philosophy of Hegel', *Interpretation*, vol. 3, nos. 2/3.

Kuhn, T.S. (1962) *The Structure of Scientific Revolutions* (Chicago).

Lenin, V.I. (n.d.) *Collected Works* (Moscow).

Lichtheim, G. (1963) 'Marx and the Asiatic Mode of Production', *St Anthony's Papers*.

Lukács, G. (1971) *History and Class Consciousness* (London).

Lukes, S. (1984) 'Can the Base Be Distinguished from the Superstructure?', in Miller and Siedentop (eds) *The Nature of Political Theory*, (London).

Luxemburg, R. (1970) *Rosa Luxemburg Speaks* (New York).

MacGregor, D. (1984) *The Communist Ideal in Hegel and Marx* (London).

McLellan, D. (1973) *Karl Marx: His Life and Thought* (London).

Marcuse, H. (1941) *Reason and Revolution* (London).

Marcuse, H. (1987) *Hegel's Ontology and the Theory of Historicity* (Cambridge, Mass.).

Marx, K. (1967) *Wage Labour and Capital* (Moscow).

Marx, K. (1971) *A Contribution to the Critique of Political Economy* (London).

Marx, K. (1973) *Grundrisse* (Harmondsworth).

Marx, K. (1976) *Capital* vol. I (Harmondsworth).

Marx, K. and Engels F. (1975–) *Collected Works* (London).

Marx, K. and Engels F. (n.d.) *Selected Correspondence* (London).

Meek, R. (1976) *Social Science and the Ignoble Savage* (Cambridge).

Merleau-Ponty, M. (1962) *Phenomenology of Perception* (London).

Moltmann, J. (1985) *God in Creation* (London).

Monod, J. (1972) *Chance and Necessity* (London).

Negri, A. (1984) *Marx beyond Marx* (South Hadley, Mass.).

Negri, A. (1988): *Revolution Retrieved* (London).

Panzieri, R. (1976) 'Surplus Value and Planning: Notes on the Reading of *Capital*', in *The Labour Process and Class Strategies*, C. S. E. Pamphlet no. 1.

Pascal, R. (1938) 'Property and Society', *Modern Quarterly*, vol. 1, no. 2.

Plamenatz, J. (1970) *German Marxism and Russian Communism* (London).

Popper, K. (1963) *Conjectures and Refutations* (London).

Poulantzas, N. (1978) *State, Power, Socialism* (London).

Prinz, A. (1969) 'Background and Ulterior Motive of Marx's "Preface" of 1859', *Journal of the History of Ideas*.

Roemer, J. (ed.) (1986) *Analytical Marxism* (Cambridge).

Rose, G. (1981) *Hegel contra Sociology* (London).

Sartre, J.-P. (n.d.) *The Problem of Method* (London).

Sartre, J.-P. (1976) *Critique of Dialectical Reason* (London).

Schmidt, A. (1969) *The Concept of Nature in Marx* (London).

Skinner, A. (1965) 'Economics and History – The Scottish Enlightenment', *Scottish Journal of Political Economy*, vol. 12.

Walsh, W. H. (1975) *Kant's Criticism of Metaphysics* (Edinburgh).

Williams, R. (1973) 'Base and Superstructure in Marxist Cultural Theory', *New Left Review*, no. 82.

Wittfogel, K. A. (1953) 'The Ruling Bureaucracy of Oriental Despotism', *Review of Politics*, July.

2

Historical-Materialist Science, Crisis and Commitment

JOSEPH FRACCHIA and CHEYNEY RYAN

Paul De Mann has suggested that the rhetoric of 'crisis' plays a central role in constituting the modernist viewpoint. For some time, it has been a characteristic of our intellectual life that people are constantly announcing 'crises', 'crises' which are claimed to be harbingers of the 'end'. In the twentieth century, this rhetoric seems to be associated with a certain cataclysmic consciousness associated with developments in late capitalism. Spengler is a founding figure in this regard. But, and this seems to be De Mann's point, the rhetoric of 'crisis' cannot be detached from that form of historicism which achieves fruition in thinkers like Hegel, and which identifies the 'crises' of birth and death – of individuals, of states, of cultures – with the progressive development towards some final end. A 'crisis' is something that occurs on the way to something else. But if we question the presumption that we are necessarily on the way to something else, then we question the conditions which give the rhetoric of 'crisis' the sense it has.[1]

The task of a post-modern viewpoint, following De Mann's view, would not be one of leaving the rhetoric of 'crisis' behind – as if the historical age could simply cast off its conditions like so much baggage; rather, initially at least, it would be to inject a little more self-consciousness, specifically a little more *irony*, into the rhetoric of 'crisis' – wherever it arises. But this is precisely what many critics of Marxism, who claim inspiration from post-modern thinking, fail to do. They talk of the 'crisis' in Marxism, and the need to 'surpass' it, oblivious to the fact that central themes in post-structuralist thought have served to render such talk deeply problematic. For example, the 'crisis' in Marxism is often attributed to its continued

46

adherence to a teleological vision of history that is theoretically naive and politically pernicious. But what are we to say of the 'crisis' of Marxism, if post-structuralism, in exposing the illusions of the teleological, has exposed the illusions of 'crisis' talk as well?

For a start, we would not *deny* that Marxism is in 'crisis'. In fact, we shall suggest as we proceed that there are reasons for believing that Marxism has always been in crisis – just as there are reasons for believing that bourgeois society has always been in crisis; sometimes it is just more obvious than others. (The two points are not unrelated, since Marxism is itself the product, perhaps the highest product, of bourgeois society.) But once the permanency of crisis is acknowledged, the question arises whether the notion of crisis, which seems to presume for its sense periods of stability to contrast with periods of rupture, any longer bears the same meaning – or any meaning at all. And this returns us to the prior point, that what is required at this time is not just, or not even, attempts to resolve the 'crisis' of Marxism so much as renewed consideration of what a 'crisis' is and means.

Self-consciousness about the meaning of crisis is further compelled by essential features of the Marxist tradition. For Marx's critique of political economy was, among other things, a sustained reflection on the meaning of crisis – in the context of the social relations of capitalist production. Remember that, at the time, bourgeois political economists, while acknowledging the empirical reality of economic crises, refused to acknowledge their importance for either the basic structure of capitalist relations, or the theory of those relations. (Say's Law – the claim that aggregate demand and aggregate production are necessarily in balance under capitalism – is perhaps the key moment in this refusal.) Hence, bourgeois political economy was by and large a sustained effort in denying any significance to crises. The oppositional literature on this matter, be it the more reactionary views of Malthus or the more progressive ones of the utopian socialists and the like, still misunderstood the significance of crises in Marx's eyes. Hence, Marx's own work, particularly his *Theories of Surplus Value*, contained both sustained and suggestive reflections on the whole question of what makes a crisis significant for theoretical and practical purposes, i.e. when does it reveal something essential and when does it reveal something purely contingent? And what Marx's example shows is that reflection on the nature of 'crises', even reflection on them as they occur in theory, is not a substitute

for examining society around us; it may rather be the most appropriate route to such examination.

We shall begin with some general remarks on the nature of
theoretical crises, inspired in part by the influential work of Thomas
Kuhn and the response to it by theorists identified with the movements of post-modernism or post-structuralism. These remarks presume that, once the postivist- and empiricist-inspired preconceptions
about the nature of theory have been undermined, the whole
question of the nature and import of 'theoretical' needs to be
rethought. Nowhere is this task more urgent than in the Marxist
tradition, where talk of crisis has become so omnipresent. We shall
then suggest one way of thinking about the Marxist theoretical
project which, by stressing its essential openness, undercuts some of
the cruder claims about the crisis of Marxism at the same time as it
reveals the more significant questions raised by Marxism's current
and problematic state. It is important to stress the unity of the more
philosophical and the more scientific approaches to the question of
crisis: Marx's own reflections on theory were deeply grounded in his
efforts in theory, and we would seek to follow his model. In
conclusion, we suggest some directions in which further theoretical
reflection on Marxism's crisis might proceed.

Kuhn and Dialectics

Thomas Kuhn's *The Structure of Scientific Revolutions* might be the
most widely-cited and least-read, anyway least-digested, book in the
history of academe.[2] At best, what social scientists seem to glean
from it is a modest critique of empiricism, i.e. that the meaning,
hence the empirical content, of statements is always relative to their
theoretical context. At worst, they simply transpose their positivist
and empiricist views of science into Kuhn's more modish jargon, so
that instead of speaking of their 'theory' they speak of their 'paradigm' – unmindful that, in Kuhn's view, the two are importantly
distinct. Kuhn's work, of course, is not without its ambivalences or
ambiguities; so any reading of it – such as ours – must be a strong
one in Harold Bloom's sense. Moreover, Kuhn's work cannot be
accepted uncritically by the Marxist tradition, if only because it
shares with other work in academic philosophy of science views on
the relation of theory and practice which are generally idealist. In
fact, as commentators like Marx Wartofsky have noted, a good deal

of what has struck philosophers of science as new and fresh in Kuhn's vision is already to be found in Hegel. This explains the comfortable fit between at least some of Kuhn's ideas and those of Marx, insofar as both adopt a dialectical approach. But the link with Hegel also suggests the importance of subjecting Kuhn's ideas to the criticisms of a more materialist viewpoint. We shall say some things about the shortcomings of Kuhn's work in this regard, mindful that the issues raised are deep and difficult ones, some of which already animate a vast philosophical literature which we cannot hope to encompass.

The Articulation of Paradigms

One way to think about Kuhn's view of crisis is that it loosens the connection between a crisis and empirical refutation or confirmation. On the old view, still held by many, a theoretical crisis is brought on by the empirical refutation of a theory's basic laws (or by the absence of empirical confirmation: the difference between these is important to the debate between Popper and Logical Positivism, but is not important here). A crisis, then, is when a theory fails to fit the facts, and it necessarily requires at least some revision in the theory – tinkering, if the failure to fit is minor, wholesale scuttling, if it is major. The first question that Kuhn raises about this picture clearly involves the notion of the 'empirical'. Kuhn maintains that the empirical evidence for a paradigm bears an internal relation to its theoretical postulates – what counts as 'empirical evidence' is always relative to the paradigm. Hence, it is naive to conceive of empirical evidence as the neutral court of appeal against which paradigms are judged. This is Kuhn's familiar questioning of the empiricist model. Less familiar is his questioning of what positivists call the deductive nomological model.

According to that view, the heart of a theory is its laws, theoretical statements to which empirical statements stand in a relation of logical deduction. If a statement thus deduced proves false, as in the case of a false prediction, then (according to this model) revision if not rejection is demanded. Now, according to Kuhn, the heart of a paradigm is a great achievement, a new and striking *result* which may alter not only our way of looking at the world, but our way of looking at theory as well.[3] Typically, the greatness of this achievement rests in its addressing problems which previous inquiry had produced but not resolved. But, just as typically, it 'addresses' these problems by redefining them, sometimes even dismissing them (hence the doubts this picture raises about the notion of theoretical progress). The first change this introduces is from thinking about laws and empirical statements as the heart of theory to thinking about problems and solutions as the heart of a paradigm. This might seem a minor one (it is anticipated in the work of Popper); but it involves an insistence on the primacy of scientific practice insofar as it makes the most important question we ask of a paradigm, not 'What does it say?' but 'What has it done?' If a theoretical tradition has had no striking theoretical successes, and this is arguably the case for many social theories, then it has not reached the paradigm stage. There may be theory, but no paradigm – and certainly no science.

So far, it is easy to see much of Kuhn's view as echoing Hegel. The idea that theory and evidence are internally related, hence the latter cannot be 'neutral court of appeal' for the former, is an essential part of Hegel's dialectical vision which has been adopted by all those he has influenced – Dewey is a case in point. The idea that problems and solutions, rather than laws and empirical statements, constitute the heart of scientific achievement is also important to Hegel and his followers – Collingwood is a case in point. Finally, the notion that science is inaugurated by an achievement, whose articulation is then the task of those working under its influence, is an element in Hegel's view that reflection, and philosophical reflection in particular, is always done after the fact – the 'Owl of Minerva flies at dusk'. This is one place where Hegel's view comes importantly into conflict with that of Marx, who ascribed a more revolutionary role to theory.

The articulation of the paradigm involves on one level what Kuhn calls 'normal science', an activity he associates with 'puzzle solving'. The term 'puzzle' is the key one here. Once a paradigm is inaugurated, the empirical work which is done under its influence generally has the character of our activity in, say, solving the Sunday crossword puzzle. It is an activity of 'filling out' what already exists in outline, an outline which not only directs us in where to proceed, but also provides us with confidence that solutions are to be found; an activity which, on the whole, is ill-conceived as one of testing, either by confirming or refuting, the paradigm. Now it is this account of normal science as puzzle solving which Kuhn's critics see as marking a deeply conservative dimension in his view. For it suggests that most scientific activity is not very self-critical (compare this with the Popperian view that self-criticism is, in some sense, the essence of science). It is probably unfair to Kuhn's insights, however, to put too much weight on the 'puzzle' metaphor, with its quietist implications, and not attend to the further possibilities opened by the notion of empirical work as *articulation*. We shall explore the value of this notion when we turn to Marx.

This notion of science as the articulation of a paradigm raises the question of *methodology*. The positivist conception of science tends to conceive of method, specifically the empirical and critical method, as independent of the particular results it generates, somehow floating above actual scientific practice while valorising that practice *as* scientific. To think scientifically, according to this conception, just means following the right method, and (it is assumed) if one employs that method properly, one is assured of achieving scientific results. Now one aspect of Kuhn's view is that methodologies are, to a significant extent, *internal* to paradigms: methodologies will vary from paradigm to paradigm, for a paradigm is partially constituted by its conception of how science should proceed. Different paradigms will differ in their conceptions of what it means to do science. But these conceptions themselves require filling out, hence another element in the articulation of a paradigm is the articulation of its methodological approach – an approach which, though implicit in the achievement which founds the paradigm, does not precede that achievement in any fully developed form.

Two points of particular importance follow from this. According to this view, the process of paradigm articulation involves both empirical inquiry and methodological self-reflection, and, as the work of Paul Feyerabend has done much to show, these elements are not always distinguishable from one another. Against the positivist view, then, that science is one thing, philosophy another, and that the task of philosophy is among other things to clarify their difference, the Kuhnian view tends to blur this distinction. Hence it has contributed to the emergence of a philosophically informed history of science, of a sort that positivism tended to discount.[4] With its blurring of the science–philosophy distinction, the effect of the Kuhnian view is to cast doubt on the importance of a question which seemed of enormous importance to positivism – the question of what constitutes the essence of 'science' or the 'scientific'. Logical empiricists, with their concern for 'cognitive significance', Popperians with their concern for the 'criteria of demarcation' – these philosophers and others felt that the question of what is scientific and what is not was an absolutely crucial question to be resolved. Today, this question does not occupy anything like the attention it once did among philosophers of science, in part because the narrative it presumed – of science's 'liberation' from religion and metaphysics – is no longer accorded the same credibility.

The internal relation of methodology to paradigm is of enormous importance for understanding the nature of criticism and the conditions which give rise to a paradigm's crisis. If one takes 'methodology' to include basic notions as to what a theory seeks to achieve and the conditions of its success, the Kuhnian view suggests that the terms by which a paradigm is assessed are set by the paradigm itself; hence, to assess the health of a paradigm, one cannot just import criteria of adequacy from the outside, one must look into the paradigm itself – to grasp its own criteria. This is another element in Kuhn's view which leads some to accuse him of endorsing dogmatism, for it raises the spectre of a paradigm whose own criteria of success render it immune to criticism. In fact, the view we have described is little else than the Hegelian view that theories, like everything else, stand or fall according to their *internal* contradictions, i.e. they succeed or fail according to criteria of success which they propose for themselves.

Marx and Theoretical Crisis

Before addressing, still in general terms, the question of theoretical crisis, let us note some implications of the foregoing for approaching Marx.

One relates to the emphasis on achievement. The heart of Marx's work is clearly *Capital*. To recognise Marx's achievement there as the centre of the Marxist paradigm, in Kuhn's sense, is necessarily to adopt a more modest view of Marxism's theoretical reach than is often taken. This can be seen by considering the question of Marxism's relevance to understanding socialist societies. Certainly, Marxism's general views on the nature of history, contained in his historical materialism, have implications for the nature and development of post-capitalist societies. Historical materialism, that is, provides a basis for developing a theory of socialist societies, but it does not itself constitute such a theory. The question of whether Marxism provides a scientific analysis of such societies stands or falls, on the Kuhnian view, on the question of whether Marxism provides us with any striking theoretical achievements in the understanding of such societies. We think that it is, at best, arguable whether Marxism has provided such insights into the working of socialist orders (the fact that Marxism provides us with such insights into the workings of capitalist society does not ensure that the Marxist viewpoint can be projected on to other realms). Our conclusion, though, is not that the Marxist project has thereby failed in its endeavours, but that its endeavours need to be conceived more modestly.

A further point involves the earlier remarks on methodology. Insofar as Marxism does provide an analysis of capitalist society, the terms for assessing that analysis are internal to the Marxist paradigm. Again, the point is not to shield Marxism from criticism; it is to insist, rather, that any criticism of a crisis in Marxism must begin with a consideration of Marx's methodological reflections, for it is there that we identify, in identifying what Marx sought to achieve, the terms of criticism for assessing that achievement. This is, in fact, how Marx himself proceeded in criticising his opponents. His criticisms of the classical political economists, beginning with the 1844 *Manuscripts*, were always careful to take them on 'their own terms', begin with 'their premises', to determine how well they succeeded by their own notions of success. The contrasting approach, which just appeals to some external standard or goal in assessing its object, is one which Marx identified with utopianism in political practice, and religion in philosophy – for both of these constituted forms of criticism which stood outside their object rather than working through it.

Kuhn's discussion of how paradigms enter into a crisis stage is interwoven with his discussion of *anomalies* – failures of a paradigm to succeed according to its own standards, failures which bear a strong resemblance to the dialectical notion of contradiction. As anomalies pile up, a paradigm may enter into a critical phase. But Kuhn strongly insists that a paradigm is never given up simply because it faces significant empirical or theoretical problems. A paradigm is never foresaken until a better one presents itself, so that the question of crisis is always a *relative* one: assessing the depth of a paradigm's crisis means attending to the alternatives available. If there is no alternative, then it is rational to maintain commitment to the existing paradigm – whatever its problems. 'As in political revolutions, so in paradigm choice – there is no standard higher than the assent of the relevant community.'[5] Kuhn's picture is one in which a single paradigm commands the general consent of the scientific community and prevails until another one arises to supplant it. But whatever the accuracy of this picture for the natural sciences,

for the human sciences it seems unrealistic to expect such paradig-
matic victory, the universal acceptance of any one theory. Precisely
because the 'relevant community' may be split along political lines,
no paradigm in the human sciences will attain the position of
dominance characteristic of natural science. We should expect, in
other words, a greater prevalence of contradiction in the human
sciences, a situation akin to permanent crisis, insofar as there will
always be more than one sharply competing paradigm as long as the
conditions of class division persist. But this makes Kuhn's point
about the comparative nature of paradigm judgement all the more
important: the depth of Marxism's crisis can only be assessed relative
to its alternatives, and the sort of frustrations which they face.

Even where the frustrations of a paradigm are clear, and alternatives to the paradigm present themselves, Kuhn is reluctant to
describe the opting for one paradigm over another as a matter of
choice. This is undoubtedly the sharpest point where problems of
dogmatism and irrationality present themselves to Kuhn's vision, for
he talks as if the adoption of a paradigm were not a matter of rational
assessment. It is in this context that Roy Bhaskar has spoken of
Kuhn's 'super-idealism'.[6] Hence, it is here that Kuhn's ideas seem to
dovetail with the irrationalism that many see lurking in much post-
structuralist thinking. There is, however, a way of framing the
problem of paradigm-adoption as Kuhn conceives it which not only
illuminates the role of rationality, but also paves the way for
considering the place of political considerations within such adoption.
It is to regard our relation to a paradigm as a matter of *commitment*.

The relation between science and commitment was a prominent
theme in the work of Micheal Polanyi, which preceeded and in many
ways anticipated Kuhn's views.[7] Following Polanyi, we might say
that our relation to a paradigm is not a matter of belief *about*
something, but a matter of belief *in* something. It is analagous, say,
to belief in another person, in the personal realm. The point about
all of these is that, while they can be rational or irrational, the logic
which they possess is ill-conceived on the simple-minded models of
empiricism and its progeny – models which identify the logic of
acceptance solely with the logic of testing a conjecture to which we
stand in an external relation. To think in terms of commitment is to
think in terms of an internal relation to the object of commitment:
when we are committed to another person, or when we are committed to a paradigm, the commitment can define our identity in such a
way that changing our commitments can have the character of a

'conversion' (in Kuhn's words). The point is not that the logic of such conversion is totally arbitrary, but rather that its logic is deeper than ordinary epistemology allows. Some read Kuhn's ambiguities in characterising that logic as evidence that none really exists, but we prefer to read them as presenting a problem to us – to explore more carefully than has been explored the logic of such conversion, say when it occurs in social theory.

Such an endeavour is one aspect of the reassessment of crisis, which we have recommended above. In the case of social theory, posing the problem as one of commitment helps us attend to the political dimensions of paradigm-adoption by attending to the relation of theoretical commitment to the other sorts of commitments that define us as social subjects. We can regard commitment, in other words, as the relation which mediates theory and practice. So that Marxist theory is regarded as one element in the articulation of broader commitments, the commitment to the proletariat (or the proletariat's commitment to itself), and the practice to which it gives rise – the political, not just the theoretical, practice – is in turn regarded as the articulation of commitment to that theory. The question of crisis, then, is posed within this nexus of commitments; reassessing the meaning of crisis means reassessing, as undoubtedly must be reassessed, the nature and meaning of the commitments which constitute the Marxist project.

Given the weight placed on commitment, it is worth noting the complete absence of reflection on this aspect in the writings of post-structuralist thinkers. Our suggestion, of course, is that along with expressing their flight from politics, this blocks consideration of the deeper questions of rationality involved in theory-adoption, since those are best theorised in terms of commitment. It would be interesting to explore how this inattention to commitment evolved, since commitment was so much the concern of existentialist thinking. Undoubtedly, at least for French philosophy, it involves the problematic place which Merleau-Ponty occupies in the genesis of post-structuralist thinking, beginning with the eclipse of his vision of Marxism by the Althusserian vision.

Theory and Praxis

The foregoing considerations have suggested that the consideration of Marxism's crisis must begin with a consideration of Marxism's theoretical aspirations – for only in their light can the theoretical project's success or failure be adequately assessed. The problem with much discussion of Marxism's crisis, we believe, is that it fails to acknowledge the open-ended character of the Marxist project. That open-endedness reflects, we shall claim, a certain tension between theory and its object which is constitutive of the Marxist project; indeed, we shall suggest why that tension may even be theorized as a permanent crisis within Marxist theory – raising questions, of the sort we have already noted, about the conditions of crisis talk in general. The open-ended character of Marxism is particularly evident when its philosophical dimensions – i.e. its methodological self-reflections – are not detached from the other dimensions that are normally identified with its social theory proper. Hence, our account will involve a sketch of how the various dimensions of the Marxist project relate to one another.

Insofar as our goal is to understand the *praxis* of Marx's 'historical materialist science' (*Wissenschaft*), a word is in order on the meaning of 'science' in this context. In the Kuhnian view, it is important not to make a fetish of science, and what it is and is not. Marx himself, in his many remarks on science throughout his writings, provides no clear demarcation as to what (in his mind) constitutes the scientific and what does not. As Althusser rightly noted, any conception of science must be constructed from Marx's achievement; we would question, again on Kuhnian lines, whether the construction of such a concept has the importance that positivists have attributed to it. But one dimension of a scientific theory, as Marx conceives it, is that it contains within it an account of its own achievement: how it arises from the social order it studies, and how it is adequate or inadequate to that study. Methodological self-reflection, then, becomes a hallmark of a truly scientific theory, and this supports the general view which we are advancing – that to assess a theory, we must pay constant attention to the theory's professed aspirations.

The Openness of Marx

In his early philosophical writings, Marx presents himself as having achieved an *Aufhebung* of the Western philosophical tradition, one which involved a materialist reconceptualisation of the subject-object relation. Marx's views here are familiar. Our purpose in rehearsing them is to draw out their implications for Marx's mature scientific endeavours – in particular, the epistemological problems which they pose for those endeavours. As Marx saw it, the philosophical tradition had distinguished human beings from animals on the basis of their capacity to reason, and had identified the uniquely human subject with the knowing subject and its object with the object of knowledge. This view relegated the bodily, material dimension of human life to at best a subordinate position in the philosophical analysis of human existence. Influenced both by the naturalism of Feuerbach and the activism of Hegel's dialectic, Marx reconceived the human subject as a sensuous-acting (*sinnlich-tätig*), producing subject, and its object as the object of that action – the natural, material world. With this turn, Marx not only 'experienced . . . the embarrassment of having to take part in discussions of so-called material interests', but he also had built the foundation for his next 'very important' undertaking, 'a work polemicising against German philosophy' which would precede his 'positive construction'.[8]

The work which Marx was anticipating here was, of course, the *German Ideology*, which he retrospectively called a 'settling of accounts' (*Abrechnung*) with his philosophical past and a successful completion of his primary purpose, 'self-clarification'. That clarification brought further changes in his thinking. While the *Economic-Philosophical Manuscripts* of 1844 presented Marx's materialist redefinition of the subject-object relation, it still retained both the Feuerbachian conceptual framework of species-essence (*Gattungswesen*) and alienation (*Entfremdung*), along with the Feuerbachian method of exposing the doubling (*Verdoppelung*) of the world. This method led to the powerful critique of the stunted growth of human beings in capitalist society which Marx performed in the *Manuscripts*, but he came to regard it as static, *a priori*, and ultimately ahistorical insofar as it could neither explain the origins of alienation nor provide any historical account of the possibilities for overcoming it. Thus in the *German Ideology* Marx sought to develop a materialist conception of history from his materialist understanding, and this meant further changes in the philosophy itself.

The *German Ideology* marked a number of key changes in Marx's approach to class society. He cast aside the more essentialist Feuerbachian conception of species-essence for a most historical conception of human nature as the 'ensemble of social relations', and he turned his attention from alienation to exploitation as the key to what is most inhuman in the class order.[9] But this deeper sociological vision constituted a 'settling of accounts' with philosophy insofar as it situated theory, hence philosophy, within the larger and ongoing 'life process' of society; in exposing the social horizon of intellectual labour, Marx questioned the independence and the primacy which much philosophy had claimed for its own activity. This has implications not only for how we study the activity of theorising, but also for how a theorising which is methodologically self-reflective understands itself.

To situate consciousness, hence theory, in the context of both material needs and historical development is to raise serious questions about the capacity of theory – such as Marx's own theory – to achieve the sort of final and complete understanding which Hegelian philosophy had sought. To begin with, Marx's own anti-utopianism committed him to working within the present and using the resources of the present to understand and change it. But the need for radical change was premised on his conviction as to the limited and partial nature of the present – the insufficiency of its social forms for realising our deepest aspirations, which includes the limited nature of its theoretical forms for achieving full understanding. So Marx found himself in the dilemma of all true revolutionaries, of having to employ tools and resources in coping with the present which his own commitments implied were deeply limited. Behind this dilemma lay a further one, rooted in Marx's more philosophical views. Hegel could acknowledge the historical dimension in all theory, but still make claims for the completeness of his own theory because he claimed to stand at the end of history, i.e. in an essentially post-revolutionary situation. Marx, though, professed to stand very much within the process of ongoing revolution. Moreover, the revolution which he envisioned would not put an end to history, rather it would constitute the *beginning* of true history. Marxism, we might say, rather than having constituted the end of our theoretical labour, constitutes a moment in its beginning. But this implies, again, that Marxism must contain a good deal of irony as to the completeness of its own theoretical endeavours.

The fact that Marxism's account of reflection seems to imply limits on its own reflections is one which many philosophers within the Western Marxist tradition have recognised. On the whole, they have sought to transcend those limits by identifying some privileged postion for Marxist reflection. This is how we understand Althusser's rejection of 'historicism', in *Reading Capital*, and Lukács's appeals, in *History and Class Consciousness* and other works, to an idealised proletarian consciousness, whose agent is the vanguard party. But there is another option, which involves (to employ a characteristic move by Marx) regarding what seems to be a problem as a solution. In that option, the limits of Marxist reflection are acknowledged and accepted, as bespeaking of the *openness* of Marxism to further transformation and change. Marxism acknowledges its partiality, in a manner we shall discuss in a moment, and in so doing foresakes adopting an overly rigid conception of its own achievement – which will induce talk of crisis, at the slightest crack.

A Modest Abstraction

We shall now explore how Marx's response to the epistemological dilemma just outlined is to recognise the historical limits of thought, hence to construct his science of the social order on a *very modest* epistemological foundation. Although he agrees with Hegel that the 'correct, the scientific mode of presentation' is the ascent from the simplest to the most concrete category, Marx's critique of the power of the concept results in a quite different epistemological orientation. For Hegel, the structure of concepts which science provides is taken to capture the essence of reality. Intellectual labour is required to fill out the structure, but the Hegelian notion that dialectical thought always returns to its starting point, that it is necessarily circular in this sense, means that such intellectual labour does not alter the initial structure; it merely subsumes reality under such concepts.

Marx's materialist critique, while acknowledging that concept is the indispensable tool of intellectual labour, implies a refusal to reduce reality to concepts. Marx insists that a conceptual presentation, such as the presentation of the capitalist mode of production in *Capital*, is necessarily abstract – and incomplete. It does not, indeed cannot, include everything relevant to understanding bourgeois societies in their particular, historical forms. For if a proper understanding of the *capitalist mode of production* can be gained only through theoretical abstraction, a complete understanding of *bourgeois societies* requires empirical analysis of that which does not fit the conceptual model. In regarding the conceptual presentation, in its abstractness, as only the first stage in a much longer climb toward concrete historical reality, Marx assumes a quite different attitude than Hegel towards that presentation. For Hegel the task of integrating empirical data into it is at most simply a matter of 'filling in the details'; if the data could not be subsumed under concepts, the truth which Hegel ascribed to concepts licensed him to ignore such data as merely 'contingent'.[10] Marx's more modest attitude toward the concept leads him to ascribe much greater importance to the 'contingent'. For the 'contingent', while it may not have a place in the abstract conceptual analysis, still possesses a truth by virtue of its historical reality, and it must necessarily be incorporated in a full presentation of the social object in its concreteness.

In this manner, Marx's conceptual analysis constitutes an 'abstract presentation of the essential'.[11] It *temporarily* defines as contingent, and *temporarily* abstracts from, real existing elements which do not fit into the conceptual presentation in order to construct a model of how the capitalist mode of production is structured and how it functions. Methodologically, this demand to reintegrate, rather than exclude or subsume, the 'contingent' not only requires forever renewed empirical analysis, it implies a constant dialectical movement between theory and empirical analysis. Its materialist perspective on the concept and its object recognises an inescapable epistemological tension between them: theory is necessary to put empirical data in a meaningful context, and empirical analysis is necessary to correct the abstraction of theory. Both are necessary, each alone is necessarily insufficient. The praxis of Marx's historical science consists of the attempt to reduce the gap between the concept and its object, to move in the direction of the concrete totality. But it is assumed that such knowledge cannot be contained in a closed book, which will provide 'absolute knowledge' for the reader; rather, knowledge is an open-ended project which cannot be completed in the Hegelian sense; Marx's *Capital* must therefore be read as an open book.

It is in this spirit that we understand Marx's claim, in the *German Ideology*, that with the adoption of a truly scientific viewpoint 'philosophy as an independent branch of knowledge loses its medium of existence'.[12] If we identify philosophy with the aspiration to universality, all that remains of that aspiration is 'a summing-up of the most general results, abstractions which arise from the observation of the historical development of people'.[13] And Marx immediately adds: 'Viewed apart from the study of real history, these statements are only *abstractions* which have in themselves *no value whatsoever*' (our emphasis).[14] Their only remaining function is to 'serve to facilitate the arrangement of historical material, to indicate the sequence of its separate strata. But they by no means afford a recipe or schema, as does philosophy, for neatly trimming the epochs of history'.[15] Marx ascribes to such abstractions the important albeit limited status of 'guiding threads' (*Leitfaden*), to orient empirical research.[16] The general theory of history in the *German Ideology*, and the summary of the dynamics of historical change in the Introduction to the *Critique of Political Economy*, may be read as presentations of these threads. It is essential that the vision they present not be detached from the epistemological views which limit its reach, for otherwise that vision is transformed into a universal philosophy of history – whose pretensions far exceed its capacities

The Crisis of Paradigms

In the methodological remarks of the *German Ideology* just noted, Marx is mindful of the obstacles standing in the way of developing an adequate theory of society: 'our difficulties begin only when we set about the observation and arrangement – the real depiction – of our historical material, whether of a past epoch or of the present. The removal of these difficulties is governed by premises . . . which only *the study of the actual life-process of the individuals of each epoch will make evident.*'[17] The difficulties Marx has in mind are not explicitly named, but it is clear that he is assuming that the ways by which the tension between concept and its object will be addressed through empirical analysis will themselves become clear only in the process of that analysis. Hence, the initial studies for *Capital*, the *Grundrisse* and the *Critique of Political Economy*, and *Capital* itself are not just excavations of the capitalist order, they are experiments in historical materalist methodolgy.

If, then, *Captital* is Marx's methodological paradigm, it is, in contrast to Kuhn's insistence that an established paradigm may be taken for granted,[18] a tentative, because incomplete, paradigm. It is therefore necessary to determine precisely the relation between *Capital* as paradigm and the historically existing capitalist societies. There are two elements of this relation, which, taken together, also establish the necessity and the nature of historical materialist science as the means of articulating the paradigm. On the one hand, although *Capital* abstracts from concrete capitalist societies, it does immediately contribute to the historical delineation of those societies. As Helmut Reichelt and Alfred Schmidt have shown, while Marx in *Capital* was not attempting to write the history of the capitalist mode of production, the logical sequence of the categories presented in that work does trace the economic prerequisites and therewith the historicity of the capitalist mode of production.[19] Futhermore, the chapter on 'Original Accumulation' establishes the social prerequisites of the capitalist mode of production. On the other hand, however, beyond determining these economic and social prerequisites, and despite the constant employment of historical examples, *Capital* is still not a historical analysis. As an 'abstract presentation of the essential', it is an *ahistorical* analysis that provides the premises – but definitely not the conclusions – for a historical analysis of the capitalist mode of production. Given the limited nature of Marx's aspirations with regard to history, let us consider the further abstractions Marx made in his conceptual account of capitalism.

To develop that account, Marx abstracted from at least four elements of bourgeois societies that need to be reintegrated in the process of moving beyond the abstract to the concrete. (1) Marx abstracted from national variations in the evolution of capitalism because he saw *Capital* as a general study of the 'capitalist mode of production and the conditions of production and exchange corresponding to that mode'.[20] While Marx took his historical examples from England as the classic or 'pure' case of capitalism, believing that all Western and Central European nations would go through the same general development, he assumed that there would be variations which would be essential to understanding the particular nature of the class struggle in the different nations. Such particularist understanding of the various national histories is the only basis on which a concrete and effective political tactic may be developed. (2) Marx abstracted from real human beings for purposes of the conceptual presentation. As he states in the Preface to the first volume of *Capital*: 'I paint the capitalist and the landlord in no sense *couleur de rose*. But here individuals are dealt with only in so far as they are the personifications of economic categories, embodiments of particular class-relations and class interests.'[21] This was not a reduction of the real historical individuals to bearers of economc categories, but a methodological signpost that *Capital* is concerned only with the basic structure and tendencies of the capitalist mode of production and exchange in its pure, 'uninhabited' form. (3) Marx abstracted from all classes in the society except those essential to the capitalist mode of production – the bourgeoisie and the proletariat. The lingering power of the nobility and the anxious volatility of the threatened petite bourgeoisie are irrelevant in *Capital*, though essential to understanding the concrete constellation of political power in the various bourgeois societies. (4) Marx abstracted from the intervention of what he termed 'superstructural' elements in the working of the economy. For example, though fully aware of the importance and complexity of the relationship between state and economy – of the ways that state intervention could mitigate the laws of laissez-faire capitalism, or could soften the class struggle by instituting forms of protection for workers – Marx did not analyse that relationship.

Through abstractions such as these, Marx focused on the structure and logic of capitalist production as an exploitative mode of production. But the three volumes of *Capital* do not make any concrete statements about capitalist societies in their historical diversity, nor (we maintain) do they make any iron predictions about their allegedly inevitable evolution. The conceptual analysis of *Capital* represents only a first but necessary step in Marx's scientific project.

The next step in that project is the move to the concrete presentation of the various, really existing bourgeois societies. In several letters Marx and Engels both protested against those who simply substituted the historical materialist theory of history and the abstract paradigm of capitalism for the concrete study of history.[22] Marx's celebrated denial of being a Marxist was made in response to the 'fatal friends' of the materialist conception of history who handled the theory so casually. The point is crucial not only theoretically, but politically. For, in Marx's view, it is impossible to develop an effective political tactic without a concrete understanding of the particular situation in each country, the unique constellation of forces and the diverse forms of hegemony (to use Gramsci's term). Any simple minded derivation of political tactics and goals from the abstract account of *Capital* alone must lead to an abstracted and irrelevant, even dangerous political tactic.[23] Engels provided a clear warning concerning this in responding to a query about his opinion of political developments in Russia: 'Marx's historical theory is, in my opinion, the essential condition of every *coherent* and *consequential* revolutionary tactic [*die Grundbedingung jeder* zusammenhängenden *und* konsequenten *revolutionären Taktik*]; in order to ascertain this tactic, one need only apply the theory to the economic and political relations of the country in question.' Lest this be interpreted as simply forcing a preconceived grid onto various and diverse countries, Engels added; 'But in order to do this [apply the theory], one must know the relations; and as far as I am concerned, I know too little of the contemporary situation in Russia to assume that I have the competence to judge the details of the tactics which are required there at a given moment.'[24] The elasticity of Marx's method thus has its counterpart in the elasticity of political praxis according to the particular situation in each country. This is one reason why Marx wrote into the rules of the First International that it should serve as an international coordinating committee, but that each national organization had to develop its own tactics according to its own circumstances.[25]

Marx and Kuhn

The move from the abstract to the concrete called for here would seem to constitute the articulation of the paradigm, the practice of normal science, as Kuhn conceives it. The achievement, in this case, is *Capital*, but it is an achievement which, when conjoined with the methodological self-reflections which accompany it, very much recognizes its own limits. On one level, empirical research articulates the paradigm by revealing its full meaning – hence 'value', in Marx's words. On another level, empirical research, by taking Marx's original, abstract model to its limits, reveals what these limits are.

Now in Kuhn's work, the suggestion is that the paradigm remains constant until it is overturned by a 'scientific revolution'. There seems to be little room for self-criticism, and self-transformation, in normal science. But in Marx's project, as we conceive it, there is more room for a dialectical relation of mutual correction between the abstract and the concrete. This is because Marxism, in contrast to the more general Kuhnian vision, contains a quite specific account of how social theory arises from its object, an account which implies some definite limits in theory's capacity to comprehend its object. This is how we understand Engels's reference, in a review of Marx's *Contribution to the Critique of Political Economy*, to the theoretical categories of the conceptual presentation as 'corrected mirror-images' (*korrigierte Spiegelbilder*) of the historical reality, images which must be modified according to the laws which the real course of history provides (*korrigiert nach Gesetzen, die der wirkliche geschichtliche Verlauf selbst an die Hand gibt*).[26] Historical analyses serve to correct or update the purely conceptual analysis – a fine example of this is found in Antonio Negri's chapter, included in this volume.

Kuhn refers to the 'essential tension' involved in scientific research at a time of scientific crisis, when the traditional paradigm seems to be losing its explanatory power but is not yet replaced by a new one. In Marxism, however, the tension is not a temporary phenomenon which appears in times of crisis. Indeed, there is reason to think that Marxism is in a permanent state of crisis – in the following sense.

The tension which defines the Marxist theoretical project might be generally described as that between the abstract and the concrete. Stated in such general terms, though, it is easy to see the affinities between this tension and that which stands at the heart of capitalist society, for Marx's political economy theorises this tension at that between the abstract (abstract labour; or exchange–value) and the concrete (concrete labour; or use-value); it claims that capitalist society continually tries to reduce the concrete to the abstract, and its failure to do so defines the historical limits of its system, the locus of its penchant for crisis. We have earlier commented on Marxism's inherence in the order it indicts, and would transcend. We are now suggesting that Marxism's immersion in its own times may even extend to the *logic* that animates the central tensions in its project. Marx said the same thing, of course, about the proletarait (whose science Marxism aspires to be). Marx characterised the proletariat as embodying the contradiction between universal and particular which informed all of class society. What distinguished the proletariat was its awareness of this contradiction, and of the impossibilities of transcending it within the limits of class society. The same might be said for Marxist theory, that it is, or should be, distinguished by its awareness of the tensions which its historical position introduces into its project.

Placing this tension at the heart of the Marxist project might lead us to regard Marxism, like the society whence it arises, as always in crisis, or it might lead us – as we suggested at the outset – to reconsider the meaning and import of crises, as we understand them. In any event, the notion of science as the project of articulating a paradigm also provides a straightforward means of responding to the post-modernist critique of Marx's theory as just another in a long line of 'essentialist' philosophies.[27] If there is one common thread to the post-modernist critique of Marx (which it also shares with positivist 'modernists'), it is the refusal to see Marx's historical materialist science as an open-ended project and the resulting attempt to 'freeze' that project at the level of its 'paradigm' and thereby to reduce it to an 'historical-philosophic theory of history' – exactly that which Marx expressly denied and, as the foregoing analysis has tried to show, avoided.[28]

Let us conclude with some words on political practice. In the history of Marxism there have been tragic (this term is not an exaggeration) consequences of conceiving of *Capital* as a closed book. The belief that Marx said all there was to say about capitalism – that normal Marxist science consisted simply of puzzle solving or 'filling in the details' – has led both to a passive and to a dangerously elitist notion of the unity of theory and praxis. The former is best represented by the *revolutionäre Attentismus* of the Second International;[29] the latter, though it has had many forms, is embodied in both Lenin's conception, taken from Kautsky, of the necessarily limited, trade-union consciousness of the working class and Lukács's subsequent insistence on the reified character of the proletariat's 'empirical consciousness'; both result in an avant-gardist definition of the party as the bearer of class-consciousness. The costs of this sort of arrogance, which defines the unity of theory and praxis hierarchically as the dominance of theory over a praxis, are evidenced in recent events in the Soviet Union and Eastern Europe. From this point of view, then, the source of crises in Marxism is an inflation of the claims of theory. Since, however, Marx's conception of the unity of theory and praxis defines both as open-ended projects, that which the theoreticians of crisis might perceive as a crippling anomaly is actually the next task.

Notes

1. Thus Paul Feyerabend, who most thoroughly rejects a progressivist vision of science, also rejects the importance of 'crisis' talk. See *Against Method* (London, 1975: rev. edition, 1988). The same scepticism about the rhetoric of 'crisis' leads Derrida, in *Positions* (trans. by Alan Bass, London, 1987), to question talk of 'epistemological breaks'.
2. Thomas Kuhn, *The Structure of Scientific Revolutions* (Chicago, 1962). Our discussion of Kuhn will draw on the book as a whole.
3. Ibid., p. xiii.
4. This, too, echoes Hegel; hence the same thoughts are found in French philosophers of science, like Bachelard and Canguilhem, influenced by the Hegelian tradition.
5. Kuhn, *Structure of Scientific Revolutions*, p. 94.
6. Roy Bhaskar, *Reclaiming Reality*, (London, (1989), p. 11.
7. See Micheal Polanyi, *Personal Knowledge* (Chicago, 1958).
8. Marx, Preface to the *Critique of Political Economy*, in R. Tucker (ed), *Marx–Engels Reader* (hereafter: MER) (New York, 1978), p. 3; Marx,

letter to Karl Leske, in *Marx–Engels Werke* (Berlin (East)) (hereafter: MEW), XXVII, p. 448.

9. Marx, *German Ideology*, MER, p. 145. That is not to say, of course, that Marx had no notion of any constants in human existence. See Perry Anderson, Discussion, in *Marxism and the Interpretation of Culture* (Urbana, 1988), p. 334.

10. As Hegel says in the *Naturphilosophie* (*Werke* (Frankfurt a.M., 1970), IX, p. 35), since nature consists of chance, arbitrariness, and disorder, ‘it is the greatest impropriety (*das Ungehörigste*) to demand of the concept that it should comprehend such contingencies [*Zufälligkeiten*]’. This is, for Hegel, true not only of nature, but also of such natural ‘contingencies’ in the social world as who owns how much. (See *Rechtsphilosophie*, *Werke*, VII, p. 112.

11. Kosmas Psychopedis, *Gesellschaftswissenschaftliche Begründung und Historische Reflexion* (Göttingen, 1981), p. 219.

12. *German Ideology*, MER, p. 155.

13. Ibid.

14. Ibid.

15. Ibid.

16. Preface to the *Critique of Political Economy*, MER, p. 4.

17. *German Ideology*, p. 155 (our emphasis),

18. Kuhn, *Structure of Scientific Revolutions*, p. 19.

19. See Helmut Reichelt, *Zur logischen Struktur des Kapitalbegriffs bei Karl Marx* (Frankfurt, 1973); Alfred Schmidt, ‘*Über Geschichte und Geschichtsschreibung in der materialistischen Dialektik*’, in *Folgen einer Theorie* (Frankfurt, 1972) and *Geschichte und Struktur* (Munich, 1973).

20. Marx, Preface to *Capital*, I, in MER, p. 295.

21. Ibid., p. 297.

22. See Marx’s letter to N. K. Mikhailovsky, November, 1877, in *Marx–Engels: Selected Correspondence* (Moscow, 1955), p. 293, in which he protests against the attempt to turn *Capital* into a ‘historico-philosophic theory’; Engels’s letter to Conrad Schmidt, 5 August, 1890, in ibid., p. 393, which explains Marx’s claim that he was not a Marxist.

23. This is, for example, the problem of Lukács’s derivation of ‘ascribed forms of consciousness’ as the basis of his theory of organization and political strategy.

24. Engels, Letter to Vera Sassulitsch, 23 April, 1885, in MEW, XXXVI, p. 304.

25. Karl Marx, *On the First International*, The Karl Marx Library, I (New York, 1973), p. 15.

26. Friedrich Engels, *Karl Marx, ‘Zur Kritik der politischen Oekonomie’*, in MEW, XIII, p. 475.

27. See, for example, F. R. Ankersmit’s recent characterization of all approaches to history before post-modernism as ‘essentialist’, in ‘Histo-

riography and Post–Modernism', in *History and Theory*, XXIII, no. 2 (1989).
28. Marx's letter to Mikhailovsky (see n. 22).
29. See Dieter Groh, *Negative Integration und Revolutionäre Attentismus* (Frankfurt, 1973).

3

Interpretation of the Class Situation Today: Methodological Aspects

ANTONIO NEGRI

Introductory Note to Theses 1 to 3

These first three theses take up the conclusions of my previous works on the theory of value. In English see Revolution Retrieved *(London, 1988) and* Marx Beyond Marx *(South Hadley, Mass., 1984). In Italian see* La forma Stato *(Milan, 1977) and* Macchina tempo *(Milan, 1982). These first three theses have methodological importance. Those who are not familiar with my previous works cited above may find them difficult. In this case, I would recommend reading the text beginning at thesis four and returning to these first three theses at the end. In the above-cited works I have continually sought to bring two traditional thematics – (1) the question of the validity of the law of value and (2) the development of the transition between socialism and communism – into contact with the new phase of political history: (3) the subsumption of the entire society in the process of capitalist accumulation and therefore (4) the end of the centrality of the factory working class as the site of the emergence of revolutionary subjectivity. In these first three theses I want to affirm the principle that the contemporary end of the economic function of the law of value, inasmuch as it is tied to a previous and outdated organization of labour and accumulation, does not diminish the centrality of the contradictions tied to social labour. The new subversive political subjectivity, then, is to be identified on this new terrain.*

Thesis 1: By Constitution I Understand the Socio-Political Mechanism Determined by the Law of Value

The form of value is the material representation of the organisation of collective labour in a determinate society. When we say 'representation' we mean that the form of value is a conceptual product. But when we say 'material' representation we mean something different, namely that the value-form, in addition to being a *representation* of the social constitution, also *corresponds* to the social constitution; or, more precisely, it is inscribed on the structure of productive cooperation and of the system of distribution and reproduction of value produced in a determinate society. The 'mode of production', or the system of production of a society, resides, in a manner of speaking, at the 'base' of the form of value; the latter, instead, constitutes the socially effective and representative mediation of the labour processes, of the norms of consumption, of the models of regulation – it resides, in short, 'above' the mode of production. The mode of production is the form of value without the representation of the social constitution. The form of value is instead the transcendental material of a determinate society – it has, then, a higher ontological intensity than the simple mode of production.

The form of value is defined by the critique of labour. The critique of labour comprises two elements: firstly, the analysis of labour; secondly, the critique proper. Now, the analysis of labour is neither simply an analysis of political economy, nor simply an analysis of ideology, law and the state; it is an analysis of all this gathered under the category of the political. The analysis of labour is therefore an analysis of the politics, or more precisely of the constitution, of a determinate society. But the constitution is the mechanism of the labour of a multitude of subjects, and therefore the product of the determinate functioning of the law of labour-value. Here, consequently, the analysis of labour becomes the critique of labour. And where the analysis of labour shows that the development of social labour produces either a process of accumulation of value or a complex of norms of distribution, the critique of labour breaks this synthesis, unhinges this constitution and marks the singularity and the dynamism of the antagonisms which the form of value comprehends.

The rules of solution of the antagonisms which are fixed by a constitution are historically modifiable. The form of value is always the result of a relation which changes according to the historical movements of a society. But since the historical changes are determined by the development and the level of solution of the antagonisms, we can say that the form of value is a function of the antagonisms and a product of their solution. The form of value, as the material transcendental of the constitution of a multitude, is submitted to the alternatives which the social antagonisms determine: it can therefore alternatively tend either toward identifying itself with the 'mode of production' or, on the contrary, toward being critically lived through revolutionary practice.

In Marx's *Capital*, volume 1, part 1, the form of value is presented to us as (1) a form of equivalence, and therefore as the form of a relation, (2) a relation whose constituent parts are historically determined, and then (3) as the dynamic of an exchange relation; which (4) moves toward a maximum of abstraction and (5) in this movement exposes a mystery (value as equivalence) which (6) hides the antagonistic character of the relation, of its form, of the corresponding mode of production. This first series of Marxian definitions of the form of value are synchronic, but in them a diachronic gash already begins to open: in points 2 and 4, for example, since it is clear that the historical determination of the antagonism and the definition of its dynamism demand an ontological indentification, a subjective ground, the materialisation of the tendency. It is perfectly logical, then, that in the third and fourth parts of volume 1 of *Capital*, Marx exclusively adopts the diachronic discourse: the analysis of the form of value here becomes historico-political discourse, in which historical modification integrates the theoretical definition and the materiality of the ontological fabric fixes the possibility of praxis.

The limit of Marx's consideration consists in the fact of reducing the form of value to an objective measure. This forced him, against his own critical premises and against the wealth of his own analysis, to consider the historical development of capital according to linear tendencies of accumulation and, consequently, it prevented him from successfully showing the movements of class struggle in light of catastrophe and innovation. Historical materialism, even in prophetic texts such as the *Grundrisse*, runs the risk of constituting a natural history of the progressive subsumption of labour under capital and of illustrating the form of value in the progressive, albeit utopian, deterministic process of perfecting its mechanisms.

Thesis 2: Even Though the Law of Value is in Crisis, Labour is the Basis of Every Constitution

When we say that there is a crisis of the law of value, we mean that today value cannot be reduced to an objective measure. But the incommensurability of value does not eliminate labour as its basis. This fact becomes clear when seen from a historical perspective.

When Marx speaks of a 'mode of production' he unfolds a history of the world which sees the passage from an Asiatic culture to a medieval mode of production to a bourgeois and capitalist mode of production. Within this last stage Marx defines the different phases of the history of the labouring process, from simple cooperation to manufacture to large-scale industry. It seems important here to assume this second series as an appropriate definition of the 'mode of production'. Today, in effect, the 'mode of production' represented by large-scale industry and its development envelops, and makes a function of its own interests, not only the bourgeois capitalist mode of production, but also the socialist capitalist mode of production and every residue of the others. When the capitalist process of production has attained such a high level of development so as to comprehend every smallest fraction of social production, one can speak, in Marxian terms, of a 'real subsumption' of society in capital. The contemporary 'mode of production' is this 'subsumption'.

What is the 'form of value' of the 'mode of production' which is called the 'real subsumption'? It is a form in which there is an immediate translatability between the social forces of production and the relations of production themselves. In other words, the mode of production has become so flexible that it can be effectively confused with the movements of the productive forces, that is, with the movements of all the subjects which participate in production. It is the entirety of these relations which constitutes the form of value of the real subsumption. We can develop this concept affirming that this form of value is the very 'communication' which develops among productive forces.

If 'communication' constitutes the fabric of production and the substance of the form of value, if capital has become therefore so permeable that it can filter every relation through the material thicknesses of production, if the labouring processes extend equally as far as the social extends, what then are the consequences that we can draw with respect to the law of value? The first and fundamental consequence is that there is no possibility of anchoring a theory of measure on something extraneous to the universality of exchange. The second consequence is that there is no longer any sense in a theory of measure with respect to the immeasurable quality of social accumulation. In the third place, even the space for the development of the labouring relations, the productive routes within society, the interactions among labouring subjects, all this is also – by definition – immeasurable.

But the immeasurability of the figures of value does not deny the fact that labour is at the basis of any constitution of society. In fact, it is not possible to imagine (let alone describe) production, wealth and civilisation if they cannot be traced back to an accumulation of labour. That this accumulation has no measure, nor (perhaps) rationality, does not diminish the fact that its content, its foundation, its functioning is labour. The intellectual and scientific forces which have gradually become central in production are none the less powers of labour. The growing immateriality does not eliminate the creative function of labour, but rather exalts it in its abstraction and its productivity. The substance of value is more important than the forms which this may assume, and it is posed beyond the very division (which is now being eclipsed) between manual labour and intellectual labour. The abstract is more true than the concrete. On the other hand, only the creativity of labour (living labour in the power of its expression) is commensurate with the dimension of value.

Thesis 3: Exploitation is the Production of the Time of Domination Against the Time of Liberation

If the law of value were to consist simply in the definition of the measure of labour, then its crisis would imply the crisis of the capitalist constitution of society. But since the law of value cannot be reduced to the definition of measure, and since it still affirms, even in its crisis, the valorising function of labour, and thus capital's necessity to exploit it, we must therefore define what this exploitation consists of.

The concept of exploitation cannot be made transparent if it is defined in relation to the quantity of labour extorted: in fact, lacking a theory of measure, it is no longer possible to define these quantities. In addition, it is difficult to make the concept of exploitation transparent if we persist in separating, dividing, searching for transcendences or solid points internal to the circulation of social production, of communication as the pervasive mode of production.

The concept of exploitation can be defined only if it is counterposed to the processes of subsumption in their totality. From this point of view, the concept and the reality of exploitation can be recognised within the nexus which links political constitution and social constitution. It is in fact the political constitution which overdetermines the organisation of social labour, imposing its reproduction according to lines of inequality and hierarchy. Exploitation is the production of political lines of the overdetermination of social production. This is not to say that the economic aspect of exploitation can be negated: on the contrary, exploitation is precisely the seizure, the centralisation and the expropriation of the form and the product of social cooperation, and therefore it is an economic determination in a very meaningful way – but its form is political.

In other terms, the concept of exploitation can be made transparent when it is considered that in mature capitalist society (be it bourgeois or socialist) a political extortion of the product and the form of social cooperation is determined. Exploitation is politically produced as a function of capitalist power from which descends a social hierarchy, that is, a system of matrixes and limits adequate to the reproduction of the system. Politics is presented as a mystification of the social process, and therefore as a mechanism which serves at times for use, at times for neutralisation and at times for blocking the processes of the socialisation of production and labour. In the period of the 'real subsumption', the political tends entirely to absorb the economic and to define it as separate only insofar as it fixes its rules of domination. Therefore, the separateness of the economic, and principally of exploitation, is a mystification of the political, that is, of who has power.*

The law of value considers labour as time in which human creative energy is unfolded. In the political constitution of advanced capitalism, the fundamental function of power is that of stripping from the social process of productive cooperation the command over its own functioning – of closing social productive power within the griddings of the system of power. The time of power is, therefore, the exploitation of social time, in the sense that a machine is predisposed to emptying out the meaning of its liberatory goals. Exploitation is therefore the production of an armoury of instruments for the control of the time of social cooperation. The labour-time of full, whole social cooperation is here submitted to the law of the maintenance of domination. The time of liberation, which is the very time of the highest productivity, is therefore cancelled in the time of domination.

Introductory Note to Theses 4 to 10

In these theses I consider post-Fordism as the principal condition of the new social organisation of labour and the new model of accumulation, and post-modernism as the capitalist ideology adequate to this new mode of production. I call these two conditions together the real subsumption of society within capital. In these theses my task is to define the economico-political contradictions of post-Fordism and to demystify post-modernism. In Politics of Subversion *(Cambridge, 1989) I tried to give a complete account of this development.*

Thesis 4: The Periodisation of Capitalist Development Shows that We are at the Beginning of a New Epoch

Here we are interested in that period of the industrial revolution which, from the years around 1848, Marx describes as the period of 'large-scale industry'. Marx also studies the preceding period of 'manufacture' – the origins of which are based in the epoch of 'primitive accumulation' and the construction of the modern state – but his interest is focused specifically on the latter period. The arc of the development of 'large-scale industry', described by Marx in its origins in the central capitalist countries, stretches well beyond the horizon of his scientific experience – it lasts, in fact, for more than a century longer, up until the revolution of 1968.

We can here summarily describe this great period of the industrial revolution emphasising principally the fact that it is divided in two phases and that this division is situated around World War 1, 1914–1918.

The first phase of 'large-scale industry' extends, then, from 1848 to 1914. It can be characterised in the following ways. (1) From the point of view of labour processes: the worker is for the first time treated within the command of machinery and becomes an appendage of the machinery itself. The labour force, here attached to the productive cycle, is qualified (this is the period of the 'professional worker'), with a clear knowledge of the labour cycles. With respect to the previous period, of 'manufacture', the technical composition of the working class is now profoundly changed because the artisan is thrown into the factory and the worker's qualification, formerly independent, becomes here instead the prosthesis of machinery which is always more massified and complex. (2) From the point of view of the norm of consumption: this first phase is characterised by a continually greater affirmation of mass production regulated only by the capacity of capital to produce it and not commensurate with an adequate wage capacity, with a corresponding effective demand – therefore, it is regulated by the determination of a profound irregularity of the economic cycle with frequent, catastrophic falls. (3) From the point of view of models of regulation: the state is developed towards ever more rigid levels of institutional integration between the construction of financial capital, the consolidation of monopolies and imperialistic development. (4) From the point of view of the political composition of the proletariat: this phase witnesses the formation of workers' parties, based on a dual organisation (with a mass component and an avant-garde component, a syndicalist and a political) and on a programme of workers' management of industrial production and social organisation, according to a project of the socialist emancipation of the masses. Here, the technical composition of the professional worker finds an adequate translation in the political composition of socialist organisation. The values of labour and the capacity of productive factory labour to dominate and give meaning to every other activity and social stratification are assumed as fundamental.

The second phase of the period of 'large-scale industry' extends from World War 1 to the revolution of 1968. It can be characterised as follows. (1) From the point of view of labouring processes: there is a new technical composition of the proletariat, and that is a type of labour force made completely abstract with respect to the industrial activity to which it is attached and, as such, the labour force is reorganised by Taylorism. Great masses of workers, who are thus 'dequalified', are inserted in labour processes which are both extremely alienating and complex. The 'mass worker' loses the knowledge of the cycle. (2) From the point of view of the norms of consumption: this is the phase in which Fordism is constituted, namely a conception of the wage as an anticipation of the acquisition of goods produced by mass industry. (3) From the point of view of norms of regulation: little by little, encouraged by Keynesian politics (but also, in general, by the reflection on the cyclical crises of the preceding phase), the model of the interventionist state comes to be formed, to support productive activity through the maintenance of full employment and the guarantee of social assistance. (4) From the point of view of the political composition of the proletariat: while the experiences of the socialist workers' organisations continue (it is principally the Soviet experience which perpetuates the bankrupt political hegemony of the old figures – the 'professional worker', now transformed into the stakhanovite Soviet superman!), new forms of organisation are configured, primarily in the USA and in the most advanced capitalist countries. In these forms of organisation of the 'mass worker' the avant-garde acts within the mass level, developing the great conceptual rallying points, such as 'the refusal of work' and 'wage equality', radically refusing every form of delegation and reappropriating power in mass and base forms.

Clearly, these two phases are made unified and distinct by the level of the continually increasing intensity of the domination of industrial capital over the entire society. The division of the first from the second phase of this period is marked by the passage to a higher phase of the abstraction of labour, or, more precisely, by the passage from the hegemony of the 'professional worker' to the 'mass worker'. We are now at the beginning of a new epoch. The tendency toward an always greater abstraction of labour has in effect disappeared and new, original and radical perspectives of development have appeared.

The new epoch starts in the years immediately following 1968. It is characterised by the fact that: (1) the labouring processes are always more radically conditioned by the automatisation of the factories and by the computerisation of society. Immediately productive labour loses its centrality in the process of production, while the 'social worker' (and that is the complex of functions of labouring cooperation transported into the social productive networks) assumes a hegemonic position. (2) The norms of consumption are once again led back to the choices of the market, and from this point of view a new type of individualism (founded on the necessary presumption of the social organisation of production and communication) has the means to express itself. (3) The models of regulation are extended along multinational lines and the regulation passes through monetary dimensions which cover the world market to a continually greater extent. (4) The composition of the proletariat is social, as is also the territory where it resides; it is completely abstract, immaterial, intellectual, from the point of view of the substance of labour; it is mobile and polyvalent from the point of view of its form.

In summary, what does it mean to us that we are at the beginning of a new epoch, and no longer simply within the phase of the completion of the process of abstraction of labour? This observation means that, while in the period of 'manufacture', and more significantly in the two phases of the period of 'large-scale industry', the development of the abstraction of labour and the formation of the processes of social cooperation of the productive forces were consequences of the development of the industrial and political capitalist machine – now, cooperation is posed *prior* to the capitalist machine, as a condition independent of industry. The third period of the capitalist mode of production, after 'manufacture' and 'large-scale industry', after the phases of the 'professional worker' and of the 'mass worker', is presented as the period of the 'social worker' – which vindicates the real mass autonomy, the real capacity of collective auto-valorisation with respect to capital. Is this a third industrial revolution or the time of the transition to communism?

Thesis 5: Marx's Theory of Value is Tied to the Origins of the Industrial Revolution

The definition of the form of value which we find in Karl Marx's *Capital* is completely internal to what we have called the first phase of the second industrial revolution (the period 1848–1914). But the theory of value, formulated by Ricardo and developed by Marx, is in effect formed in the previous period, of 'manufacture', that is, in the first industrial revolution. This is the source of the theory's great shortcomings, its ambiguities, its phenomenological holes and the limited plasticity of its concepts. Actually, the historical limits of this theory are also the limits of its validity – notwithstanding Marx's efforts, at times extreme, to give the theory of value the vigour of a tendency.

To make our discussion more specific, let us note that already in the course of the second industrial revolution, and in particular when we find the passage from the 'professional worker' to the 'mass worker', essential characteristics of the theory of value begin to fade away. The distinction between 'simple labour' and 'socially necessary labour' loses every importance (except that of continually stimulating absurd arguments), showing the impossibility of defining the genealogy of 'socially necessary labour'; and most importantly, the distinctions between 'productive labour' and 'unproductive labour', between 'production' and 'circulation', between 'simple labour' and 'complex labour' are all toppled. With regard to the first couple, already in the second phase of the second industrial revolution, but continually more so as we enter the third industrial revolution, we witness a complete dislocation of the concepts: in effect, 'productive labour' is no longer 'that which directly produces capital', but that which reproduces society – from this point of view, the separation from 'unproductive labour' is completely dislocated. With regard to the second couple, it is necessary to recognise that 'production' is 'subsumed within circulation', and *vice versa*, to a continually greater extent. The mode of production finds in circulation its own form. With regard to the third distinction, also in this case we witness a complete redefinition of the relationship between 'simple labour' and 'complex labour' (or qualified or specialised or theoretical or scientific labour). It does not become a linear relation which can be led back to a quantity but rather it is an interaction between completely original ontological stratifications.

Finally, the criteria of exploitation come to be placed under critique. Its concept can no longer be brought back within the category of quantity. Exploitation, instead, is the political sign of domination above and against the human valorisation of the historical/natural world, it is command above and against productive social cooperation. Now, even though this definition of exploitation is certainly contained within the intent of Marx's philosophy, it is none the less not clearly expressed within the historical limits of his theory.

Thesis 6: The Laws Constitutive of the Form of Value are the Laws of its Deconstruction

The processes of the transformation of the form of value, the passages from one period of capitalist development to the next, follow the dynamic of the capitalist social relation and they are determined by the antagonistic relationship of exploitation. These processes are developed in the form of a rudimentary and effective dialectic: by exploiting the labouring forces, capital employs these forces within structures which coercively envelop them; but these structures are, in turn, either broken or remoulded by the social forces of production. The real process is the resultant of these particular tensions; the development has no logic, it is simply the precipitate of the conflict of collective wills.

(We must insist here on the fact that no teleology is given for this development. Every result is appreciable only *a posteriori*; nothing is preconceived. Historical materialism has nothing to do with dialectical materialism. When it does come about that certain presumed laws are verified – such as, for example, the law of the tendential fall of the rate of profit which, within the limits of the second industrial revolution, does effectively describe phenomena with are undoubtedly true – even in these cases, there is no *a priori*, no preceding intelligibility; there is only the *a posteriori* truth of what comes to pass.)

On these bases, it is obvious that scientific attention will have to be focused more on the discontinuities (be they ruptures or innovations) than on the continuities: in effect, the continuities are nothing other than discontinuities or ruptures which have been dominated. As for innovations, these too are nothing other than structures of domination, but more precarious ones, because the conflict, the struggle and the refusal to work have been, at their origin, stronger. These conflicts could not have been resolved if not be means of a leap forward, a paradigm shift, a qualitative transformation. Capital, however reformist it may be, never willingly passes to a subsequent or superior phase of the mode of production. In effect, capitalist innovation is always a product, a compromise or a response, in short a constraint which derives from workers' antagonism. From this point of view, capital often experiences progress as decline.

And it *is* a decline, or, better, a deconstruction. Because the more radical the innovation is, the more profound and powerful were the antagonistic proletarian forces which had determined it, and therefore the more extreme was the force which capital had to put in motion to dominate them. Every innovation is a revolution which failed – but also one which was attempted. Every innovation is the secularisation of revolution. Consequently, within the processes of socialisation of the form of value which we have described, it becomes evident that the dialectical process, which modifies the capitalist ordering and determines the sense of its innovation, attacks capitalist power and its hagemony on the socio-political transformations of society to a continually greater degree. The growing complexity of society is the growing precariousness of domination. (The philosophers who have made of social complexity a labyrinth in which the revolutionary function of the proletariat gets lost, or the hermeneuticians who make of historical complexity a maze in which mice run indefinitely – all of these are only charlatans.) In effect, the more the laws of the transformation of the value form are realised, the more they demonstrate their efficacy as forces of the deconstruction, destruction, of power. The motor force which constitutes the form of value, the antagonistic expression of the productive force of living labour, is simultaneously the motor of the deconstruction of the form of value. As long as capital had the possibility to play its own game, as long as it had other territories where it could divert the moments of destabilisation which prepared the deconstruction, the situation could be sustained by capital and by the political forces in which it is always incarnated and identified. But now, in the phase of the total subsumption of society and the complete multi-national-isation of the productive processes, what alternative does it have left? Directly, today, the innovative process destructures, deconstructs, capital. The revolution, momentarily blocked and stalled in a sequence of innovative moments, cannot be banalised. Everyone is waiting to see to what extent the malaise of capitalist civilisation is really and simply the anarchy of meaning and the emptiness of its soul.

Thesis 7: The Deconstruction of Value is the Matrix of Subjectivity and *Vice Versa*

Deconstruction is the broken line which leads across the transformations of the form of value. But who deconstructs whom? We know the object: deconstruction is the profound, implacable, irreversible deconstruction of domination; it unfolds at the same moment as the political and social form of exploitation is determined and as its innovations are manifested. But who acts within the dynamics of this antagonism? This actor is, first of all, the multitude: it is the innumerable multiplicity of powers and social knowledges, it is the web of meanings in everyday activity. We are not yet talking about a subject, because subjective characteristics cannot yet be attributed in this galaxy. Other critical passengers are probably necessary for identifying the condensation of subjectivity. Now, though, we still have a fine powder of energy particles before us, a real and true ontological fabric of the multiplicity which undergoes deconstruction. If no subjectivity is actually achieved here, there is none the less a process of the invention of subjectivity in motion, which we recognise as inherent, consubstantial with the activity of deconstruction – a genetic source of subjectivity. The phantasm of subjectivity is the potent and fundamental fabric of deconstruction.

In the orthodox Marxism of the nineteenth century, and in any case before 1968, the functions of destruction and reconstruction were separated from the act of insurrection. The immediate strategy of struggle had to articulate destabilisation and destructuration, moments of a war of movement and a war of position. However, this separation no longer works. Destruction and reconstruction live together in deconstruction. The fabric on which the antagonistic subjectivity is defined is not a tendency which looks toward a mythic future, toward a future hypostasis – on the contrary, the process of the construction of subjectivity is also a process of deconstruction. Auto-valorisation and sabotage are the double figure of one and the same object – or, better, they are the two faces of Janus, the gateway to the constitution of the subject.

This is how we understand that if deconstruction involves a phantasm and arrives at an element of subjectivity, subjectivity cannot live except by means of deconstruction. The very form of the antagonism is defined in this new, complex and articulated relationship between subjectivity and deconstruction. If, in effect, production is already completely communication, then the sense of the antagonism will have neither a place nor a time of foundation separate from communication itself. It is in the deconstruction of communication that the subject is constructed, and that the multitude finds its power.

Thesis 8: The synchronic and Diachronic Figures of the Transformation of Value Lead to Strategic Contradictions of Development

First of all I will define the terms.

(A) By synchronic figures of the form of value I understand those which Marx shows constituting themselves around 'socially necessary labour', around the illustration of its ontological consistency. It is primarily in volume II of *Capital* that we find this formulation, and principally through two concepts. The first is the the concept of the 'mediation' or the 'equalisation' of the values of the labour force precisely in the process which sees the social dimension of this constitution itself. Now, the trend of socialisation, in same moment that it constitutes collective individualities which are always more abstract and more productive, also defines them as antagonistic entities – with respect to the command which capital would like to exercise over the subjective consolidation of socially necessary labour. The second concept which Marx amply dwells on here is that of the tendency toward the unity of production and circulation, which is realised by means of the progessive integration of the movement of value: in those days, this was accomplished through the transportation networks, today through the communication networks. Now this integrative dynamic is put in the service of the definition of the antagonism on the ontological terrain – it allows us to gather the multitude together in the antagonistic polarity.

(B) By diachronic figures of the form of value I understand those which were already described at some length in Thesis 4 – later we will return to the 'professional worker', the 'mass worker' and the 'social worker' to focus even more clearly on the material contradictions which the movement of their figures determines. Here, though, I want only to define the form of their movement, in order to specify, from the beginning, that this movement has nothing deterministic about it. Observing, in fact, the transformation of the forms of value and the introduction, by means of this transformation, of a process of the continually higher abstraction and integration of labour, we can imagine a type of motor or finalistic reason of development. But to claim this, even only in the form of a dialectical synthesis, would be to mask and hide the deepening of the process's contradiction. Instead, nothing that we have experienced allows us to arrive at the rationality and the teleology of the transformations. On the contrary: in the historical development, in the succession and separation of epochs and phases of development, constantly we are presented with only the unpredictability of the mechanisms in action, with only the struggle which is always open between the unique polarities of power and knowledge. The fact that the historical development seems to follow a rhythm marked by the passage to higher forms of the socialisation of production and antagonism does not reveal any kind of destiny: it would not be correct to impose the rules of our reading on the immense variation of historical events. In fact, these processes are highly contingent, since they are swayed in the flux and marked by catastrophes, and since their tendency, which is progressive, is shown as a dissemination rather that as unilinearity. The diachronic processes of the form of value are like fireworks and, between pauses and growth, they extend on the horizon of always more complex figures. The indications which Marx gives with regard to the qualitative leaps in the diachrony of the form of values – and, in particular, in volume I of *Capital*, when he studies the formation of 'large-scale industry', and in volume III, when he analyses the recomposition of all the components of production and circulation in the constitution of the world market, or in the *Grundrisse*, when he analyses the genesis of the 'universal collective individual' – should be taken up again and verified: so, well beyond the residues of logical determinism which are sometimes identifiable in Marx, we can verify the wealth of his historical intuition which extends the antagonism (and its movements and its tendency) over the entirety of the dimensions of development.

(C) By strategic contradictions I understand those effects which, associating themselves with the synchronic and diachronic figures of development, are determined on the limit of the emergence, or undoubtedly around the emergence, of adequate subjectivity.

To make the terms more clear, allow me to propose a few examples. In the first phase of the second industrial revolution, in the phase from 1848 up until World War I, the largest contradictions (synchronic contradictions, internal to the productive cycle) are those which open between the direct labouring processes and the capitalist process of production. The 'professional worker', situated right in the middle of the labouring process and completely in control of it, also wants control of production. The vindication of worker control and management of the labouring process and of the control of the productive cycle constitutes a strategic contradiction in this phase. And we can easily see why: it is because a subjectivity and a programme are born where the synchronic determinations and the diachronic rhythms, which generally define the period, come to maturity. Around the theme of worker control and management, the multitude of 'professional workers' constructs the matrix of a revolutionary subject and develops the communist project in an 'appropriative' model.

In the second phase of the second industrial revolution, which extends from the end of World War I to the revolution of 1968, the strategic contradiction is located between productive processes and reproductive processes, or, rather, in the extreme socialisation of labour. In this case, too, we have a multitude of labouring subjects which is caught in an enormous contradiction, in the conspiracy of the synchronic figure of the form of value. In other words, here the contradiction between the massification of 'dequalified' and abstract labour, which the workers refuse, and the general rise in the level of cooperation, of the wage levels and of the quality of needs, becomes explosive. The 'mass worker' constructs, around the 'refusal of work' and the discovery of the extremely high socialisation of its labour, its own model of communism, in terms of an 'alternative' model.

Thus we reach the epoch which we are beginning to experience, the third industrial revolution. From the 1970s on, we have had the bad luck to live in the most cruel and stupid period of restructuration and repression. But in this same period we have grasped the determination of a new, extremely high, strategic contradiction – the contradiction opened by the radical productive socialisation which is in opposition to capitalist command (be it bourgeois or socialist). The key element of this passage is the dislocation of the synchronic contradictions in the form of the political, the dislocation of the objectivity of exploitation toward the structure of command. As a result, the contradiction immediately touches on the sphere of subjectivity. The contradiction itself is revealed in that particular form of subjectivity which is antagonism. A fundamental consequence derives from this: the strategic contradiction, and that is the precipitate of the synchronic and diachronic moments of the antagonism of development, are presented in a subjective, political form – communism is proposed according to the model of a 'constituent power'. After the 'appropriative' model, after the 'alternative' model, we have the 'constituent' model, which envelops the others, carrying the strategic contradiction directly toward subjectivity. 'Constituent power' gives form to social production, it envelops the social and the economic in the political, its pulls together the organisation of production and political organisation in a radically constructive way. But we will come back to this later.

At this point, we can conclude our explanation, noting how the result we have come to is nothing other than a return to what we had maintained in Thesis 7: 'the deconstruction of value is the matrix of subjectivity'. Now, we can verify that the strategic contradictions of development show, or, better, produce and institute, a new antagonistic subjectivity. All this does not come about in a deterministic way, but instead it is the fruit of a process dominated by the multitude – which exalts its own power in freedom. We can, then, conclude our demonstration in the following way.

Thesis 9: The Strategic Contradictions of Development Verify the Laws of Deconstruction

Thesis 10: The Constitutive Fabric of the Present Phase of Capitalist Development is an Enormous Node of Strategic Contradictions

The characteristics of the current period of capitalist development (the initial phase of the third industrial revolution) were constructed in the 1970s, and specifically between 1971 and 1982. On 17 August 1971 Nixon and Kissinger depart from the gold standard – this act launches a great signal of deregulation throughout the capitalist world. It is an attempt to break the pressure, the cumulative effect, which the workers' struggles in the advanced capitalist countries and the liberation struggles in the Third World have produced in the 1960s (in the final offensive of the struggle of the mass worker). In the 1970s, the capitalist Trilateral imposes its own policies against the proletarian Tricontinental of the 1960s.

What is the project which capital imposes on this moment of development?

(A) It is, first of all, the destruction of the factory, and in particular the liquidation of the hegemony of the Taylorised process of labour. The analysis of labour is deepened and its organisation becomes progressively more decentralised spatially and instead it is focused on the expropriation of social knowledges, on the capitalisation of the social labouring networks: in short, it concentrates on the exploitation of a working figure which extends well beyond the bounds of the factory. We call this figure the 'social worker'.

(B) The project also involves the computerisation of society, and in particular the productive use of communication and the transfer of the programme of control of society from the outside (the factory) to the inside (communication) of society itself. A mode of social production comes to be defined here and its fundamental characteristic is that of integrating society (that is, in Marxist terms, reproduction and circulation) in production. In the 1970s, we were able to see primarily the ugly face of this passage: the destruction of the Fordist model, of the guarantee of employment and welfare, the construction of marginalisation and the plural labour market, the intensification of the exploitation of the strata which were poorly protected, above all women and youth, and the furious mixing of different forms of exploitation, all of which had become compatible within the socialisation of productive fluxes.

(C) It is essentially relative to this mixing of exploitation, of its diverse strata, compositions, levels, that the new state-form constructs itself: it becomes a differentiated control of the productive social totality, an organic capacity-necessity of producing crises at any moment and any place. The capitalist state, in this phase of development, is a crisis-state – and only that: it is the state which plans the crisis.

(D) Finally, the capitalist project is the spread of this system of exploitation over the entire world. At this level, we witness a process of integration (vertical, between various strata of development, and horizontal, that is, universal) of all the forms of exploitation. Capitalist neo-imperialism runs through several stages in the 1970s: first a process of multinationalisation, which is continually more explicit; then, a phase of the displacement of Taylorism and Fordism toward the periphery and the installation of a crude but effective hierarchical system which is made to function on the world level; and finally, a world financial integration continually pushed forward. We must recognise here that monetarism, used within the frame of deregulation, has been forged into a fearful device of control and repression, both against the working class and against the social proletariat.

Thus we reach 1982 – the year, that is, in which the crisis of the Mexican debt (the first among many) marked the end of the 'heroic' period of the world-wide extension of new forms of deregulation and new accumulation. With the crisis of 1982 we could see the fact that, if the deregulation had functioned ferociously against the central worker, it had only partially struck the peripheral worker – rather, the world-wide spread of the mode of producion opened passageways through which the effects of decentring appeared as perverse, sometimes rebounding against capital. The thrust of the major contradictions toward the periphery of the system brought to light a series of focal points for revolt and some possibilities for revolution, in themselves perhaps irrelevant, but capable of determining shock waves, travelling through toward the centre of the system; no longer weak rings, but weak networks. Actually, the fabric of the present is an enormous node of strategic contradictions – it is like a boiling volcano, which multiplies the explosions and fluxes. The year 1982 consolidates the crisis as a permanent form of the cycle we are now entering.

Introductory Note to Theses 11 to 15

One of the fundamental problems and insoluble difficulties of the Marxist conception of the working class derives from the fact that the relationship between workers' struggles and capitalist restructuration has a dialectical development: the struggles contribute to the develop- ment, they determine it, and they can break it only when political consciousness intervenes. The workers' struggle, therefore, is always 'within' even when it is 'against' capital. In this group of theses I put forth the hypothesis that, in the current stage of the development of class struggle (of the social worker in the real subsumption), new technical *conditions of proletarian independence are determined within the* material *passages of the development and therefore, for the first time, there is the possibility of a rupture in the restructuration which is not recuperable and which is independent of the maturation of class-consciousness. My attempt to define ontological categories of subversive subjectivity against the dialectical categories of the relation- ship struggles-restructuration was systematically developed in* The Savage Anomaly: The Power of Spinoza's Metaphysics and Politics *(Minnesota, 1991; first published Milan, 1981) and principally in* Fabbriche del soggetto *(Livorno).*

Thesis 11: Today the Revolutionary Point of Contradiction is the Antagonism between Social Cooperation and Productive Command

What differentiates the present from the previous phases of devel- opment of the capitalist mode of production is the fact that produc- tive social cooperation, previously produced by capital, is now presupposed in all of its policies, or, better, it is a condition of its existence. From this point of view, the synchronic and diachronic contradictions do not result in strategic contradictions, but they are reopened by them. Consequently, the crisis does not reveal itself as a difficulty, an accident: the crisis is the very substance of the capitalist process. It follows, then, that capital can only show itself as a political subject, as a state, as power. In opposition, the social worker is the producer – the producer, prior to any commodity, of social cooperation itself.

We should explain ourselves in greater depth. In every moment of the development of the capitalist mode of production, capital has always proposed the form of cooperation. This form had to be functional with the form of exploitation, when it did not actually inhere within it. It was only on this basis that labour became productive. Likewise, in the period of primitive accumulation, when capital enveloped and constricted pre-existent labour forms to its own valorisation, it was capital which posed the form of cooperation – and this consisted in the emptying of the pre-constituted connections of the traditional labouring subjects. Now, instead, the situation has changed completely. Capital has become a hypnotising, bewitching force, a phantasm, an idol: around it revolve the radically autonomous processes of auto-valorisation, and only political power can succeed in forcing them, with the carrot or with the stick, to begin to be moulded into capitalist form. The transfer of the economic into the political, which comes about here, and in global dimensions with respect to the productive social life, is accomplished not because the economic has become a less essential determinant, but because only the political can tear the economic away from the tendency which leads it to mix with the social and realise itself in auto-valorisation. The political is forced to be the value-form of our society because the new labouring processes are founded on the refusal to work and the form of production is its crisis. The social worker's productive cooperation is the consolidation of the refusal to work, it is the social trench where the producers defend themselves from exploitation. In opposition, the political, as a form of value, contains only mystification and extreme violence.

And the frame is not significantly changed by the very high intensity of the composition of capital which is inverted onto the social in order to control it – because, actually, the more the instrumentation of production becomes abstract, going beyond the figure of mechanisation and becoming immaterial, the more this itself is implicated in the struggle which traverses the social. Automatisation participates still, in part, in the old political economy of control by means of machinery: but computers are already beyond this horizon and offer very high potentials for possible ruptures. In communication, the immateriality is total, the commodity is transparent – here the possibilities for the struggle are very great and they are ruled only by an external power. These brief examples serve only to indicate how already, also and principally on the terrain of technological advancement, and as a direct result of technology's process of perfecting itself, there exist sectors which are vulnerable, always more vulnerable, to the autonomy of social cooperation and the auto-valorisation of proleterian subjects, to the exaltation of the individual and collective micro-physics.

All this serves as proof for our thesis that the antagonism between the social cooperation of the proletariat and the political command of capital, while being given within production, is founded outside of itself, in the real movement of the social. Social cooperation not only dialectically anticipates the political and economic movement, but it pre-exists them: it announces itself as autonomous.

Thesis 12: The Struggles Precede and Prefigure Social Production and Reproduction

Here we will investigate our proposal in Thesis 11 in the particular case (that the reappropriation of cooperation, by the social proletariat, determins a series of effects on the structure of the capitalist system) is valid in general. The rhythm of the passage from one epoch of capitalist development to another is marked by the proletarian struggles. This old truth of historical materialism has been continually confirmed by the implacable movement of history and constitutes the only rational nucleus of political science. The transformations of the machinery, the restructuration, the new norms of the customs and the new arrangement of the institutions, all follow where the struggle has been – where, that is, living associative labour has been freed and has thrust forward its own autonomous project. There would be no development if the proletarian struggle, the living associative labour, were not to succeed in giving these hammering blows, that history has witnessed, against the rigidity of the world of command, against the domination of dead labour. But the proletarian workers' struggle does not function only as a pulse of energy which brings dead and accumulated labour back to life: it is also the real entrepreneur of history, because industry, nature and civilisation are constrained to modify themselves in a way which is complementary, functional, organic to the contents, to the needs, to the tendencies, to the forms of organisation of the proletarian struggles. This is the boss's curse: those who learn most from the class struggle get ahead. This paradox is the shame of the boss – the perennial spy, who borrows and represses. The proletarian struggle, the workers' struggles and now the thousands of figures of the everyday revolt of social labour have – within the order – dominated (that is, put in motion, formed, prefigured, anticipated) the epochs and the phases of capitalist civilisation, of the industrial civilisation, which we know.

Without 1848, without, that is, the 'damned days of June', when the Parisian proletarians and the artisan population came to the centre and showed the limits of bourgeois liberty, the entrance to the historic period which we call the second industrial revolution would have been unimaginable. Without 1917–1919, and, that is, once again, the workers' revolt, incarnate in the Soviet dictatorship of the professional worker and the insurrection which spread to all countries which were even slightly marked by the capitalist mode of production, the opening of the subsequent phase (which we have seen dominated by the mass worker) would have been inconceivable. It is within this process, of leaps, of great contradictions and great explosions, that the mystery of the class struggle and history itself unfold: a mystery, we mean, for those who are constrained to suffer it – in their salons and in newspapers, always defined by a deficit of politics: not certainly for the proletariat, because it holds the effective-key of the determinations, of the leaps, of the advances and retreats.

Thus we arrive at the origins of the third industrial revolution. Here too a revolution marks its beginning, 1968; but it is a strange revolution, and even it does not become immediately comprehensible. In this revolution we have an intellectual subject, an actor of social communication: but why does it have to rebel? In fact, this intellectual substance, this activity of social communication, even though it is masked by workers' labour, is nothing other than living abstract labour. This is a new figure of the proletariat. The years following 1968 have allowed us to reach the full understanding of these metamorphoses of labour and the labour force – consequently to understand why and how the orderings of capitalist power so quickly and decisively changed themselves. Once again, capital has followed this revolutionary force, it has repressed it, it has sought to close it off in new technical dimensions of production and command. To this end, capital has radically transformed the structure of the state. It has then intervened in the urban structures, in public expenditures, in the ecological, moral and cultural dimensions of life, etc. – always following and mystifying the historical passage of humanity. In any case, notwithstanding the process of perfecting capitalist domination, we still cannot see where the potential of the struggles which were constructed in the passage of 1968 will finally end up.

Thesis 13: The Clandestine Life of the Masses is Ontologically Creative

There are two ways to nullify the power of historical materialism. The first consists of reducing the class struggle to a natural history of production – this was the principal avenue of social democrats and the specific ideology of the 'appropriative phase' of the proletarian movement (1848–1914). This becomes clear when we consider that the Bolsheviks made this ideology their own, and while they were certainly not social democrats, they were profoundly tied (given the levels of Russian development) to the first phase of the second industrial revolution. The second way is that of absorbing the class struggle in the dialectical movement of capital – this was the avenue of modern reformism and the specific ideology of the 'alternative phase' of the revolutionary movement (1917–1968). Now, both these avenues nullify the power of living labour in such a way as to lead to a synthesis, which is either vulgarly materialistic or dialectical. Meanwhile, it is only the consideration of the power of living labour, and the irreducible autonomy of its fundament, that gives us an understanding of the fact that history is a living reality and that innovation is its permanent motor. The more the theory of value, or rather the measure of exploitation, becomes old and useless, the more living labour becomes a hegemonic presence and the orienting criteria of its development. It is an ineluctable movement, which continually breaks the limits of domination and pushes forward the configurations of reality.

One might object: where has this movement gone? For decades this creative force had not made an appearance. But this is not true. Really, only those who do not want to see are blind to it – and blind mice do not always make us feel pity. Instead, we must consider the subterranean, clandestine life of the modern proletariat: this is where living labour comes from, with the cooperative, communicative and innovative characteristics which today we recognise as pertaining to labour; thousands of routes of a vast, plural, extremely mobile phenomenology. The very movement and organisation which capital decrees for society, even these are produced, inasmuch as they are disfigured, by the secret life of the masses – capital gives us a police artist's rendering of the infamous massses, but scientific deconstruction has taught us to decompose this image and discover the fertile terrain of life. Living labour, then, even and primarily in its clandestine existence, accumulates and assimilates to its own being the needs of liberation of the exploited masses and makes this new being a power which is constantly more productive. Productivity is the positive transfiguration of liberation. Therefore, this long process which is cutting away at capital must come to an end: cutting away at it, just as the refusal to work has already cut from capital the power to measure the time of labour; as proletarian association has already emptied out the capitalist vocation of fixing the forms of productive cooperation; as labour, which has become intelligent through abstraction, has already torn reason away from capital. The ontology of living labour is an ontology of liberation. The efficacy of historical materialism as a theory of freedom is based on this creative materiality.

Thesis 14: The Sequences of Proletarian Power are Asymmetrical with Respect to the Sequences of Capitalist Development

The Sequences of Proletarian power not only do not correspond to capitalist development but neither are they, in the negative sense, the inverse of capitalist development. This asymmetry is an indication of the profound autonomy of the real movement from capitalist movement. The movement is free, on the side of the proletariat. So free, in fact, that it is difficult, and often impossible, to determine the behavioural laws of proletarian power. In any case it is certainly impossible to determine (not only) laws (but even uniformity) which would have a general validity, that is, beyond the limits of the single phases which we have carved out from the flow of the epochs of development. We can certainly identify uniformities, for example, like those which are verifiable in the first phase (1848–1914) of the second industrial revolution, long periods of stasis of relations between the classes, which are translated into long and bitter depressions, often exacerbated by ruinous crises. But already at this point, identifying these uniformities does not mean constructing a theory of the crisis. In fact, if we look at the second phase of the second industrial revolution (1917–1968), here the duration of the crises is reduced to a minimum (even though they are catastrophic) and the dynamism of the mass worker, different from that of the professional worker, is extremely strong. Now, we have a new period, a new experience: what is the experience of the social worker in the crisis? The only strong uniformity and tendency has been the co-presence of crisis and development, of repression and innovation

– in other words, the co-presence of opposites. This derives in effect from the final result of the preceding history: the independence of the two forces in play, of proletarian power and capitalist power.

But this, one might object, proves only the independence, not the asymmetry, of the movements. I do not believe so: in the development of the struggles there is more than the simple independence of the movements – there is, on the proletarian side, the creativity of living labour, the unpredictability of the nodes of contradiction, there is the accumulation of unmanageable ontological masses, there is the double helix which unites the deconstruction of the adversary and the construction of subjectivity. And there is much more: the ideology, organisation, armament, finance, models of production and reproduction, centralisation and democratisation, the use of legality and violence, etc. There is, finally, and this is decisive, the autonomy in the construction of the cooperative dimension of the proletariat. It seems to me that it is always possible to read the asymmetry of the sequence of proletarian power and those of capitalist development and power; but even if, in some cases, in the past epochs this condition was not given, it certainly is in the period we are now entering.

The concept of asymmetry carries with it still other effects. In the first place, it means that the autonomy of class is not recognisable if one looks for it through the categories of capital. This is obvious: a pure and simple corollary, which was already evident in the epoch of the mass worker and in the frame of the development of the alternative organisational model – and yet, it is important to emphasise this because this obvious determination takes away any last sign of scientificity from the political economy of capital. Second corollary: the history of development, of power, of capital can be described only starting from the schizophrenia which characterises it – in opposition to workers' autonomy. Within this radical asymmetry, the history of capital is recognisable as a depotentialised ontological process – leading to formalism, leading to illusionism – or maybe now only a pathology.

Thesis 15: The Capitalist Structuration of the Social is Destructive, the Proletarian Structuration is Creative

This thesis follows and completes the preceding ones projecting them forward as a tendency. Therefore, the thesis illustrated here will seem more practical than theoretical, and this tendency will be ethical. The 'inversion of praxis' here moves through ethics; the web of meaning is constructed by ethically oriented action. Thus, the analysis is put back on its feet. Ethics is the terrain of possibility, of action, of hope. It is the site where the sense of being originates. Up until here we have excavated in the system of dead labour, of capital, of power, and we have seen how, wedged into that system, there was a clandestine, subterranean, hidden motor pulsing with life – and with such efficiency! We have, in a manner of speaking, rediscovered the Marxian affirmation of living labour in today's world, when living labour is already completely separated, autonomous and positioned against every naturalistic rigidification of being. It is equally against every dialectic, even a materialistic one, even Luxemburg's conception (though it is an honest one) which proposes the contradiction of living labour in the system of capital only on the extreme 'margins' of the system. But the affirmation of living labour is not sufficient if there does not open around this the point of view of action, of construction, of decision-making. The concept of living labour in Marx is the watershed between the critique of political economy and the construction of the party – assuming living labour in this second perspective, we call it creative activity. It is a creative activity, then, which is seperate from the perspective of capital, from its science and even from the critique of its science. The science of the rational organisation of the extraction of surplus value and the construction of profit, of the optimal allocation of resources, of planning and the reduction of the universe to a balance-sheet, all this is no longer of any interest – here living labour does not serve as a critique of this, but as a source of the auto-valorisation of subjects and groups, as the creation of social cooperation and with this, through this, of the maximum of wealth and happiness. It is a radical science, because its roots are firmly planted in activity and in the evacuation of every hypocritical protest of realism.

Introductory Note to Theses 16 to 20

In these theses I attempt to go beyond the contradictions of the dialectical theory of 'within-against' and to arrive at a 'completely new' definition of proletarian subjectivity. I mean to show that the concept of revolutionary organization can be expressed in class-consciousness only if class-consciousness is radically, ontologically, autonomous. This passage identifies the theme of a new determination of the category of the possible. I do not see the possibility of communism except in terms of a radical constitution. It is clear that my debt to the constitutionalists, from the founding fathers of the bourgeois constitutions to Hannah Arendt, is only nominal. Traditional constitutionalism is in effect a school of juridical regulation, a theory of checks and balances, or, even better, of pluralism and of the equilibrium between the classes and its reproduction. Much more important here is instead my reference to the theory and practice of communist democracy and its tradition – from the Communards to the Soviets, from the International Workers of the World to the European autonomists in the 1970s. I am convinced that Lenin is not far from these positions and I sought to demonstrate this fact in Trentatrè lezioni su'Lenin, La fabbrica della strategia *(Padua, Milan, 1976).*

Thesis 16: The Passage from the Structure to the Subject is Ontological and it Excludes Formalistic or Dialectical Solutions

The formal possibility of the passage from the structure to the subject does not solve the difficulties of the historical construction of the subject; but none the less, it clarifies them. From a formal point of view, the passage from the structure to the subject, that is, the installation of the problem of organisation for proletarian liberation, seems to follow a straight line. In fact, if inventive labour had expanded to occupy all of society (this is the true definition of modernity), then – since this labour is principally creative – it reconstructs society itself, revolutionising it through a process of subjectivisation. This process should present us with no serious problems. Still, from the formal point of view, we are dealing here with an inversion of the path which has led from labour to capital. In the same space where capital has expanded to occupy all of society, we must try to recognise how living labour precedes in unravelling capital, deconstructing it, occupying its territory and constructing in its place a creative hegemony. The subject, from this point of view, is auto-valorisation.

But when we assume the inversion of praxis no longer only from the formal point of view but as a real possibility, that is, when we adopt the point of view of action and history, the enormous difficulties arise. The inversion of praxis is condemned to meet up with intransigent resistances. Its autonomy is not consumed, unfortunately, in the spontaneity, in the happiness, in the utopia of the assault on the heavens. The obstacles and the limits that it encounters are enormous. Faced with this difficulty of movement, then, the inversion of praxis often ends up badly – rigidified, blocked. The transformative intentionality (which is the dignity of the proletariat) is frustrated and tends to fold back on itself, to the exasperation of subjective power, becoming blind voluntarism: terror, reaction, the instrumental use of the old forms of repression adopted from the enemy. To construct something new? Within a revolutionary mechanism? This is merely illusion and tragedy! Certainly, all of this has happened and will happen again. From Robespierre to Stalin, from the revolts of the 1920s to those of the 1970s, we have often witnessed the desire for transformation resulting in terrorism – victorious or vanquished, conducted by the state or by small groups, it really makes no difference: in every case, it signals the blockage of revolutionary action, and it is always in the figure of a retreat, perhaps a *resentiment*, the symptom of a defeat, the desperate resistance against an adversary which is felt to be stronger. We do not want any of this. Consequently, social democracy is posed as a means of avoiding this tragedy. But we do not want this either. In effect, we think that these defeats were not inevitable – and we will try again. Our task, then, is to recognise defeat and yet not be defeated. Pessimism of the will, optimism of the intellect: this shows once again just how far we have come from the Second International.

How, then, can we propose and manage an inversion of praxis which avoids both terroristic overflowings and Stalinist paranoia? How can we propose a praxis which, on the other hand, refuses the social democratic temptation? How can we generate a revolutionary process from the bottom, constituting its movement in society? This is clearly the fundamental point. But, once again, two equally impotent lines conflict here: one is the line of those who, with extreme subjectivism, hope to make themselves masters of the social dialectic; the other is the line of those who, immersed in the masses and the cults of spontaneity, respect the social dialectic to such a degree that they become incapable of rupture and refoundation. For the first group, the decision over reality becomes mixed with violent gestures while the second group is ruled by repetitiousness and hypocrisy. In opposition, to be effective, the inversion of praxis must realistically assume reality, in all of its complexity, refusing to cede either to the 'prometheanism' of the bosses' dialectic or to the 'narcissism' of the aesthetic of spontaneity. But the critique of Machiavellism or the critique of the utopia, the refusal of Stalinism or of contractualism, does not situate us in a self-sufficient position: instead, on the contrary, the critique is damaging when it does not succeed in maintaining the problems, their consistency, even while having emptied the unilateral and mystified solutions.

Well, the passage from the structure to the subject is possible only when all the elements which had been excluded (for being partial or unilateral or mystifying) have been now recomposed together. The passage which interests us, that by which the organisation of the proletariat for liberation is constructed, will be itself a structure: a structure in which all the real elements which constitute the struggle, the organisaton and the revolutionary life of the masses will be situated – utopia as discipline, authority as the moments of the construction of consensus, mass labour and the labour of the avant-garde, auto-valorisation and auto-organisation, the destabilisation and destructuration of the enemy, the deconstruction of the adversary and the construction of autonomous institutions of counter-vailing powers, the long experience of the historical alternative and the passion of insurrection.

And all this comes about on the terrain of ontology – in other words, outside of every dialectical or formalistic metaphysics; in still other words, by means of an operation of accumulation of the complex of the activity of transformation which the subject operates on itself.

Thesis 17: The Theory of the Workers' Party Presupposed the Separation of the Political from the Social

The problem of the political organisation of the proletariat can in no way be resolved either in the organisation of a delegated representation or in the expression of an avant-garde, even if this avant-garde tries to incorporate mass dimensions. Both of these solutions have traversed the history of the workers' movement (and also the revolutionary history of the bourgeoisie) often contradicting each other, but always weaving together. This is for a good reason: both of these models, that of delegation and that of the avant-garde, presuppose the same, common, transcendental dimension, as media-tion of historical plurality. Delegated representation and the avant-garde are mechanisms of mediation. It may be the mediation of the nation, or of the bureaucratic corporation, or of the working class – in every case it is the hypostasis of a unity over the process which separates the multiplicity from the unity, society from the state. And even when the dictatorship was presented as the goal of party activity, even then representation (of the avant-garde, in this case) was a conceptual function of mediation: in the old theory, the dictatorship was nothing other than overdetermined representation.

At this point, none the less, we must go to the heart of the problem, and emphasise how the true foundation of the traditional theory of representation is not so much the 'necessary' mediation of the social, but rather the 'arbitrary' separation of the political from the social. I mean that the specific nucleus of the concept of delegation, and in general of political representation, coincides with that of the separation between what pertains to society and what pertains to the state, between the economic and the political, between trade union activity and party activity.

Without embarking on a historical critique, permit us to note here that this theory of representation, and therefore of the separation between society and the state, does not correspond in any way to the current reality of the class struggle, to the current form of value and the contemporary constitution of society. At best, that theory is the passive legacy of an epoch which has passed by. Specifically, delegated representation, through the division between the trade union and the party, was theorised and practised in the period of the Second International (which corresponds to the first phase of the second industrial revolution). Its concept is adequate to the figure and the ideological horizon of the professional worker, to its project of the external avant-garde and the emancipative, progressive and orderly teleology of labour. The theory of the mass avant-garde, of the popular and mass party, corresponds instead to the next period, of the mass worker and the second phase of the second industrial revolution. The organisational model and the model of representation are based on the alternative project – still a form for justifying the separation of the social from the political, of the trade union from the party.

Today, the elimination of these figures of representation has become evident, because we cannot in any way define a line of division between the social and the political, let alone a line of mediation which tries to transcend the material processes which run throughout reality. This disappearance of the lines of demarcation between the social and the political, between the individual and the universal, is the fundamental characteristic of the third industrial revolution; and its consequence is the elimination of traditional representation. There is only one terrain for the expression of political will – it is immediately general, abstract and universal. Clearly, our critique is not so much in oppostion to delegation (even though the communist and anarcho-syndicalist traditions, in perfect agreement here, are very sympathetic to this line), but rather it addresses the question of the ontological conditions of representation. It is, in fact, on the ontological terrain that the problem of the workers' party and democracy must be posed – with respect to an ontology which has definitively left behind every difference between the social and the political.

Thesis 18: Today the Political Invests and Radically Constitutes the Social

Power and the political are dimensions of the social. Today there is no way to see the political as autonomous and separate. However, even though it has been led back to the social, it still does not lose the characteristics which distinguish it as a legitimating and legitimated force. Finding the political in the social is not the identification of a utopian site – on the contrary, it produces a new, powerful definition of the social. It is paradoxical and instructive to note how so many years of polemics about the 'social state' have produced an incredible deepening of the social definition of the state. Social plurality is today a plural fabric of counter-powers of counter-knowledges and counter-cultures – therefore, in addition, it is also a dissemination of rationales of legitimation and the inscription of adequate relationships of force. The inversion of praxis, in the frame of the current form of value, must also therefore be exercised completely throughout the continuity of the social and the political.

There is no mythical, political point toward which the strugglers must be concentrated in order to make history explode. The explosions must be given in the everyday – however they will not because of this be any less explosive. The discourse on communism crosses planes of plural immanentism, extremely rich tensions; from this point of view, the political investment of the social reality constitutes new horizons of experience, of language and struggle. At one time, the discourse on emancipation was pointed towards a utopian objective, according to a technique of the progressive overdetermination of the development, from the social to the political, so as to be able to spill over from there back onto the social: now, this discourse, having gradually become the mystified conglomerate of every hypothesis of measure and hierarchy, founded on the separation of the political from the social, has been exhausted, leaving room for the practices of liberation.

Thesis 19: The Power of the Proletariat is a Constituent Power

The two processes, non-symmetrical but historically complementary, of the political deficit of capitalist control (or, more precisely, of the emptiness of the rationality of its action) and the proletarian displacement of the political throughout the social arrive at a moment of truth in the crisis, when the social and juridical order is clearly incapable of showing its own validity. Only one of the two parts is still capable of innovation; only proletarian power can be a constituent power.

Constituent power has always been defined as an 'extraordinary' power with respect to the ordinary legitimacy of the constitution. This extraordinary quality consists in the fact that, in contrast to the normal situation, constituent power is able to act in ontological terms – constituent power is a legislative power which orders reality in a new way, creating institutions and normative logics. Constituent power constructs society, carrying (in the tradition) the political into the social – even if only in an extraordinary manner. This is the source of the enormous wealth of the appearance of constituent power in constitutional history: it breaks with the institutional routine, putting society and the state in communion, inventing society. But, from the humanistic and Renaissance revolution to the English Revolution, from the American Revolution to the French and the Russian, and to all the other revolutions of the twentieth century, once the extraordinary moment of innovation has ended, the constituent power has exhausted its effects. Now, this exhaustion is symptomatic of a fact, which is fundamental in the mystification that bourgeois political theory and constitutionalism have operated on the concept of constituent power: the real problem has never been that of the truth of the transition, but rather of the modality of its Thermidor. The investment which the political made in the social had to have an end: this end was the limit offered by the necessity of the production and reproduction of the victorious form of value. This is taken so far that, playing on the 'extraordinary quality', the juridical dogmatic has assumed the Thermidor in the very definition of constituent power. In short, constituent power is subjected to the same destiny as the concept of representation: where the power of representation is subjected to limits of space, and is forced to empty itself, deterritorialise itself, constituent power is subjected to temporal limits, restricted by the extraordinary quality which is the fragility of the project and its execution.

But now the conditions of the revolution of the mode of production and the perspectives which open from the inversion of existing praxis lend a strong concept of constituent power to the formation. It gives us a constituent power which is not closed, which is not limited to exceptional periods of its effects, which acts always and everywhere within the world of the institutions – which therefore also is approximated by the life world. This continuity of constituent power has actually been implicated in every constituent experience, but as a regulative idea: the Machiavellian idea of the 'return to principles', the tables of fundamental rights, permanent revolution, cultural revolution, etc. – these implicit elements have never been made reality. But today everything is different. In the new conditions of the mode of production, just as in those of the form of value, we can glimpse a repetition of experience which, without being formally a constituent power, move as if they were, with a maximum of plasticity and continuity. Constituent power is becoming an element of the life world. Through its appearance in the everyday, it loses the monstrous aspect which the bourgeoisie has given it: the indeceny of a submergence in reality, of overflowing the institutions – in other words, it escapes the necessity that it be an 'extraordinary' power. It assumes instead an entrepreneurial figure – clearly a political, public, collective function, but one which is animated by an uncontainable projectivity and an absolute plasticity.

It is important to emphasise this relationship which constituent power discovers with this entrepreneurial function. On one hand, this nexus attributes juridical power with those formal characterisitics which belong to the social and economic experience – mobility, dynamics and inventiveness; on the other hnd, this nexus shows that today, principally at the level of constituent power, it is impossible to make any separation between the political and the economic, any duplication of the functions, and it makes clear that today the political constitution can exist only as an investment of the social.

Constituent power must give the political the pregnancy of the economic, and the economic the universality of the political. Communism transforms the general management of society through the everyday activity which the multitude exercises: constituent power continually forms and reforms the fabric of these new relations of production – a political, economic, social production.

If we ever want to speak of a party again, in the third industrial revolution, it is obvious that we can do so only in terms of constituent power.

Thesis 20: Today the Constitution of Communism is Mature

When we consider the investment which the political makes of the social, so that the one cannot be said without the other, we are renouncing Max Weber's and Vladimir Ilich Lenin's use of the sociology of power. In effect, the isolation of power is no longer possible. But neither is it possible to deduce from this that society considers itself a sufficient communicative exchange. The political, while it diffuses its form over the social, is filled by the social with its content; consequently, through the political this content is raised to a constitutive form. Social exchange is primary only in this sense, in the play of the conditions which today form the political: precisely in the sense which this alludes to power, it assumes power as a condition, it is disposed in that direction. Because this is the case, a series of material conditions must be given – material conditions which allow that the political form and the social content coincide. This can come about only if the political imposes on the social the form of the most absolute equality. If there is not equality, absolute equality in the conditions of social exchange, there is not the possibility of putting power at the service of the multitude, and, that is, unifying the form of the political and the contents of the social.

Today this possibility is available to the proletariat – and only to the proletariat, because it is clear that such a relation between the form of the political and the social contents in itself eliminates, through the machine of production and reproduction of equality, the very definition of capital, be it that of the capitalist market or that of social planning. On the other hand, it is obvious that all the conditions of the current mode of production push toward the complete socialisation of political power and, conversely, toward the complete politicisation of the social. Now, these tendencies can find a way out of the mystifications, where they are now held captive, only by developing communism. Communism is today the only possible constitution in relation to the development of the mode of production and in the necessity of its unveiling. Outside of communist constitution there is no other form of value, but only disvalue and death.

Appendix on Some of the Theoretical Sources of These Theses

I would like here to point out some of the scientific positions, both philosophical and economic, which I have had contact with in recent years and in relation to which my opinions have developed. I refer first of all to the school of regulation. When this still constituted a militant school of thought in the 1970s, my relationship was that of an amicable exchange. From this point of view, the schemata used in Thesis 4 are themselves a fruit of this cultural exchange. The modulation of the theme 'workers struggles–capitalist development' and its structural development around the categories of the 'technical development' and the 'political composition' of the working class were initially elaborated by Mario Tronti in Opeai e capitale *(Turin, 1966) and by Romano Alquati in* Sulla Fiat ed altri scritti *(Milan, 1975); only later were these themes adopted by the French (Boyer, Lipietz, Coriat). In recent years the positions of the school of regulation have changed dramatically: it has become an academic school and it has attempted, with great clarity, to change the schema which workerist Marxism considered a chapter of the critique of political economy into functionalist and programmatic schemata. This does not mean that the regulation school is no longer socialist: rather, the school is defined, from the economic point of view, as an explicit reformism – while, from the philosophical point of view, it gives greater and greater prominence to voluntaristic and evaluative options – which, in my opinion, the regulation school derives from its reading of the political neo-Kantianism of André Gorz. Classical workerism is thus pushed toward the economic objectivism of the 'process without a subject' and blessed with the holy water of socialist (and today, increasingly, ecologist) good intentions. In the second place, I keep in mind above all in these theses the philosophical positions expressed in post-Sartrian French thought (Foucault, Deleuze, Guattari). Whatever the differences among these authors may be, it is clear that the critique of power (potere) constructed by them touches the law of value and tries to inscribe, in its crisis, the* beyond *of the dialectic between movements and restructuration, as a unitary mechanism, as the pressing demand to take away every truth value from the enunciations of power. In these French theories, ontology is posed against the dialectic and the possibility of ruling the relationship between social struggles and capitalist (social, productive, state) restructuration is taken away from power. The limitation of these theories consists in the fact that they pose the critique of power as a line of flight, as the splendour of the event and of the multitude, and they refuse to identify a* constitutive power *which would be the organ of the subversive minority. There are none the less indications that this thought can go beyond its current limits. Specifically, Foucault elaborated (and Deleuze developed) the evolution of three great paradigms of power: that of sovereignty, which in our language (see Thesis 4) he would present as the period from primitive accumulation to the first industrial revolution; then, the disciplinary paradigm (in our terms: the second industrial revolution); and, finally, the paradigm of communication – which, in our terminology, we situate after 1968 as the pre-eminent definition of the age of post-Fordism. According to Foucault and Deleuze, around this final paradigm there is determined a qualitative leap which allows thinking a new, radically new, order of possibility: communism. If in the society of sovereignty democracy is republican, if in the disciplinary society democracy is socialist, then in the society of communication democracy cannot but be communist. Historically, the passage which is determined between disciplinary society and the society of communication is the final possible dialectical passage. Afterwards, the ontological constitution cannot but be the product of the multitude of free individuals – this condition is possible on the basis of an adequate material structure and a process of liberation which courses throughout all of society. This group of propositions which leads the experiences of the militancy of the 1960s and 1970s to the level of highest philosophical abstraction seems to me to represent a useful and productive element of dialogue.*

---

* *Editor's note.* For a clarification of the sense in which Negri, here and henceforward, uses 'power', see: M. Hardt, translator's foreword to A. Negri *The Savage Anomaly*, University of Minnesota Press, Minneapolis/Oxford 1991, p. xi–xii, where power as 'might' (*potestas*) and as capacity (*potentia*) are distinguished.

4

The Inversion of Class Perspective in Marxian Theory: From Valorisation to Self-Valorisation*

HARRY CLEAVER

Different theories provide insights into various aspects of the social relationships of capitalism from different points of view. In Marxian terms, the usefulness of any particular theory depends upon understanding the particular *class perspective* from which it grasps those relationships. A given theory may express any of several different capitalist or working-class points of view. We do not have to agree with a theory to understand with which aspects of the class relation it is preoccupied, how it approaches them and, therefore, the ways we may find it useful. In this essay my principal concern is an examination of the usefulness of some recent work – both within and without Marxian theory – on the postive content of working class struggle, that is to say, on the various ways in which people have sought to move beyond mere resistance to capitalism toward the self-construction of alternative ways of being. As a methodological prelude to that examination I discuss, first, the issue of class perspective as applied to economic theory and, second, the question of the inversion of class perspective within Marxian theory with an example taken from post-World War II Italian Marxism.

* My thanks to Conrad Herold, Massimo De Angelis and the editors of this volume for helpful comments on this essay.

Theory and Class Perspective

For example, the class perspective of neo-classical economic theory is fairly obvious. It has been developed to maximise its usefulness to the managers and apologists of capitalism for the purposes of prediction, manipulation and legitimisation. Once we recognise this, it makes it easy to understand both the preoccupation of neo-classical microeconomic theory with markets, with the decision-making of firms, consumers and workers as well as the particular way the theory deals with those aspects of capitalism. The neo-classical theory of the firm has been elaborated specifically to provide not only an understanding of the processes and results of profit-making behaviour, but also guidelines for the maximisation of profit e.g. the equation of marginal costs and marginal revenues or the setting of factor prices equal to their marginal productivity. At the same time, the theory is constructed in such a way as to legitimate such behaviour by hiding the antagonistic class relations of exploitation which such behaviour involves. Once we understand the purposes and methods of such a theory, we can use it to provide us with an understanding of the business and ideological practices of both the bourgeoisie and their economists. Those involved in workers' struggles against capitalism can study such theory to understand how the opposition thinks, plans and justifies its actions to others. Given that the managers of capital and their economists are, as a rule, quite serious in their attempts to maintain and extend their control, studying such neo-classical theory facilitates understanding their goals, methods and strategies – the grasp of which can help workers calculate their own actions in the class struggle.[1]

Marxian theory and its categories also provide a conceptual apparatus for understanding the social relationships within capitalism. That apparatus has proved to be useful for those involved in the struggle against capitalist society because it not only expresses, with greater clarity than any other critical theory, the precise mechanisms of domination, but also renders those mechanisms transparent, and thus easier to confront and deal with. For example, whereas neo-classical firm theory confines the concept of exploitation to a special case (that of marginal productivity exceeding the wage) Marxian theory shows how all reproducible relationships between capitalists and workers involve antagonism and exploitation, and then goes on to detail the specific mechanisms through which that exploitation is

organised (the structure of the wage, the division of labour, absolute
and relative surplus value and so on).

Capitalist managers, of course, are well aware of all the concrete
phenomena associated with these mechanisms of exploitation – they
know that they can improve profits by holding down time or piece
wages, by organising production in such a way as to pit workers
against each other, by increasing the length of the working day or by
raising productivity while limiting wage growth – but neo-classical
theory, unlike Marxian theory, hides the antagonism of these
relationships while preserving their essence and hence the servicea-
bility of the theory. For example, the equation of the wage with
marginal productivity is not presented as a rule of the thumb which
guarantees the exploitative extraction of relative surplus value, but
as merely a technical condition of efficiency required to maximise
profits.

The theory of relative surplus value and the theory of optimal
factor pricing thus express two different class perspectives on exactly
the same phenomenon. Neo-classical theory provides a decision-
making tool to managers while doing so in a way that camouflages,
even to them, the antagonisms which make that tool a weapon of
domination. Marxism, on the other hand, provides workers with a
conceptual framework which allows them to penetrate the camou-
flage, to recognise the mechanism of domination and thus to think
clearly about strategies for opposing or undermining it.

The Inversion of Class Perspective *within* Marxist Theory

A great deal of Marxian theory, however, precisely insofar as it
specifies the mechanisms of domination in such a transparent
manner, remains underdeveloped by forgetting to carry through two
kinds of analysis: first, an inversion of class perspective of a slightly
different kind than the one discussed above and, second, an analysis
of the struggles against domination.

The kind of inversion of class perspective that I have in mind is
the sort that Marx employed in his analyses of surplus-value. From
the point of view of capital, surplus-value exists primarily as 'profit',
that is to say in relation to capitalist investment. Capitalists are
primarily interested in surplus value, i.e., they judge its adequacy
not so much in terms of the absolute amount of it but in relation to
the amount of investment required to produce it, i.e. the rate of

profit, $s/(c+v)$. If the rate of profit is less than that in another sphere of activity, capitalists will tend to shift their investment, even if their current surplus-value is large in absolute terms.

From the point of view of the working class, however, the essential issues of surplus-value are elsewhere. First, the absolute amount of surplus labour time being extracted from them is of great importance because it measures one part of the life time they give up to capital. Second, for workers the relevant measure of the relative size of surplus-value is not the rate of profit but the rate of exploitation, s/v, where the time given up to capital is compared to the time expended in meeting their own needs. This working-class perspective on surplus-value is hidden behind the capitalist preoccupation with profit, both in the world of business and in the world of bourgeois economics. Both these Marxian ways of conceptualising surplus-value express, in a general way, a working-class perspective. Yet, the concept of surplus-value and the concept of surplus-value as profit clearly express different and opposed preoccupations and class interests.

When we think about working-class struggle against surplus-value, whether the struggle to shorten the working day (which cuts absolute surplus-value) or the struggle to increase the value of labour power (which cuts relative surplus-vlue) we think, first and foremost, about a drop in the rate of exploitation. When we think about capitalist efforts to expand surplus-value, we know that they think first and foremost about a rise in the rate of profit. Thus, we can see in these two moments of Marxian theory a kind of inversion of class perspective within an overall theoretical approach which seeks to understand capitalism from the point of view of the working class.

At the most general level of his theory of capitalist society, Marx rarely failed to develop both sides of his analysis – although a great many Marxists have. He almost always retained his fundamental vision of capitalism as a society of antagonistic class struggle in which two class subjects confronted each other and the effort to dominate was always met by resistance and the struggle for liberation. From the *Communist Manifesto* through the *Grundrisse* and *Capital* to his later writings, it is easy to find repeated expressions of the antagonistic opposition between a temporarily dominant capitalist subject and a struggling, potentially victorious working-class subject. This opposition was at the heart of his theory of revolution and liberation, of the possibilities of moving beyond capitalism, and it is frequently expressed in the details of his theory. Sometimes, however, in working out his theoretical understanding of these relations of domination and struggle, Marx, like so many Marxists who have followed him, became so preoccupied with understanding and laying out the mechanisms of domination that he failed to develop, at a theoretical level, the kind of duality of perspective embodied in surplus-value and surplus-value as profit. Where such elaboration is missing it has often proved both enlightening and useful to work it out.

The second, and related, failure of Marx's writings occurred where the historical analysis of domination was not complemented by the analysis of the struggles those mechanisms were designed to domi- nate. Therefore, a not inconsiderable body of his writing appears at best to be lopsidedly preoccupied with the machinations of capitalists rather than with the struggles of those workers for whom Marx was elaborating his theory. One example of such lopsidedness should suffice to illustrate the point. In the analysis of primitive accumula- tion in *Capital*, Marx's discussion of the expropriation of the means of production (chapter 27) and of the bloody legislation against the expropriated (chapter 28) dwells at length on the severity of the measures employed to achieve these ends, but barely touches on the struggles by which people resisted those measures. Yet only an analysis of the bases and depth of that resistance can both explain the severity and the pattern of the measures employed and suggest lessons for more contemporary conflicts of a similar nature.[2]

The obvious counter-example from *Capital*, one which shows how Marx did treat such conflicts in ways that analysed both sides, is his discussion in chapter 10 on the struggles over the lengthening and shortening of the working day. Here, his analysis is more fully developed both at the theoretical level and in the analysis of history. The battles over the length of the working day are analysed both in terms of workers' attempts to reduce their exploitation and capital's attempts to expand or defend their profits.[3] Unfortunately, this discussion stands almost alone in the balance it shows, at both the theoretical and historical levels, between the capitalists' efforts to dominate and the working class's struggles against that domination. As a result, those who have understood Marxian theory as a weapon to be wielded by workers in their struggle with capital – as neo- classical economic theory is a weapon in the hands of capitalists – have been faced with the need to complete Marx's analysis by working out the theoretical implications for Marxian theory (and history) of this second kind of inversion of class perspective on specific issues in order fully to grasp the theoretical and political implications of his analysis for working-class struggle.

From the 'Composition of Capital' to 'Class Composition'

One of the best examples of such an inversion was that carried out by Raniero Panzieri, Mario Tronti and others of the Italian New Left in the early 1960s with respect to Marx's concepts of technological change and the composition of capital. As laid out in detail in volume I of *Capital*, Marx's theory of technological change recast capital's own understanding – as expressed, for example, in Adam Smith's analysis of the division of labour in terms of efficiency – using his own labour theory of value to focus on a key mechanism of domination: the use of fixed capital for the domination of living labour. The central concepts of Marx's theory were threefold: the technical composition of capital, the value composition of capital and the organic composition of capital.[4] The technical composition of capital denoted the particular material configuration of plant, machinery, raw materials and labour involved in a production process. An 'increase' in the technical composition of capital occurred when any productivity-raising reorganisation of that process involved the increased use of fixed capital (by a given labour force). The usual neo-classical economic counterpart to this representation of such a change is a rise in the capital/labour ratio.[5] A change in this technical composition of capital appears as a merely technical concept denoting a reconfiguration of production.

To discuss such a change in value terms, Marx introduced the complementary concepts of the 'value' and the 'organic' compositions of capital. In both cases the plant, machinery and raw materials are aggregated by their value and appear as constant capital and the labour employed is aggregated by its value and appears as variable capital. In both cases the ratio of the two can be expressed by c/v. The distinction between the value and organic compositions lies simply in Marx's desire to distinguish between changes in c/v which are due to changes in the value of constant or variable capital unrelated to changes in the technical composition and those which are due solely to changes in the technical composition. Thus, c/v is called the 'value' composition when no reference to the technical composition is necessary or desired, but is called the 'organic' composition when such a reference is required or desired. Thus the value composition (c/v) may rise simply because of, say, a drop in the value of labour power due to good weather and an unusually productive harvest that reduces the value of bread. Marx designates as a rise in the organic composition of capital, on the other hand, only a rise in c/v which results from the introduction of new plant or machinery which raises labour productivity.[6] This concept becomes central to Marx's analysis not only of technological change but also of its long-term consequences, of what he sees as a tendency of capital to substitute controllable machinery for less controllable workers. That analysis concerns the tendency for a rise in the organic composition of capital to lead to the displacement of workers, a rise in unemployment and systemic crisis.

These concepts have, with good reason, received a great deal of attention from Marxist economists over the years, especially with respect to the issues of labour displacement and crisis. The tendency of capital to displace labour and to generate a reserve army of unemployed workers has been a generally accepted part of Marxist theory among most of its practitioners. On the other hand, preoccupation with the evolution of the organic composition of capital has been at the centre of seemingly endless debates among Marxists over the meaning and importance of the socalled tendency of the rate of profit to fall and its relationship with capitalist crisis.

Recognition of the limits of Marx's treatment of these issues, however, emerged in the midst of working-class struggles around the introduction of technological changes during the post-war modernisation of Italian industry. The violent refusal of Italian workers to accept the Italian Communist Party's (PCI) view of such modernisation as a development with which workers should cooperate led some Italian Marxists, such as Raniero Panzieri, to re-examine Marx's own analysis. Through that re-examination they rediscovered a class bias to such change which the PCI and its theoreticians had been downplaying, namely the capitalists use of machinery to control and dominate the working class.[7] Whereas the PCI had emphasised the positive benefits of such change – rising productivity and thus the possibility of rising wages – they were ignoring the way in which the rise in the organic composition of capital was being used by capital to increase exploitation – and reduce the relative strength of workers. The workers, of course, saw this very clearly and the gap between their experience and the PCI's response was enough to make them angry over the PCI's complacency before such change. The work of Panzieri and his colleagues gave a theoretical articulation to that anger.

This work led to deeper, more detailed analyses, not just of theory, but of actual changes in industrial production processes and their relationship to the issue of workers' power.[8] It was this work which led to the kind of theoretical and political inversion I want to emphasise. Whereas Marx's work focused on how a rise in the organic composition of capital was a means to the realisation of relative surplus value, these Italian theorists linked this with Marx's closely related study of the division of labour and drew new implications from Marx's own work. If Marx had recognised, as he did, how any given division of labour was always a vehicle of capitalist control, they argued, then we must also recognise that any change in that division would have an impact on the structure of that control. Moreover, Marx's own argument that technological change was often introduced in response to workers' struggles[9] could not only be reinterpreted as a response to a breakdown in the structure of control built into the existing division of labour, but could also be seen as the introduction of a new division of labour aimed at restoring control.

In this way, the focus in the study of technological change was shifted from innovation in the means of capitalist domination to the dynamics of the class struggle in which workers overcame one such mechanism and capitalists responded by trying to introduce a new one. This kind of research thus led to a new series of concepts to study technological change as a moment within the changing balance of class power: class composition, political recomposition and decomposition.[10] To Marx's concept of the 'composition of capital', these Italian theorists juxtaposed 'class composition'. Both concepts refer to the same phenomenon: the organisation of the production process, but whereas the emphasis in Marx's concept is on the aggregate domination of variable by constant capital, the concept of 'class composition' involves a disaggregated picture of the structure of class power existing within the division of labour associated with a particular organisation of constant and variable capital. Moreover, the concept of 'class power' here is associated not only with the power of capital to dominate, but also with the power of workers to resist, which is directly related to the intra-class distribution of power among workers. All divisions of labour, it was pointed out, involve some kind of hierarchical distribution of intra-class power – usually codified in a wage hierarchy. This shift to 'class composition' thus opened the door to a much more complex kind of analysis of class forces than Marxists had ever associated with the concept of 'composition of capital'. It was both a theoretical and political enrichment of Marxist theory.[11]

This is even more true for the concepts of 'political recomposition' and 'decomposition' which were developed to deal with the all important dynamics of change in technology and the division of labour. While it can be said that capital seeks a 'class composition', i.e. a particular distribution of inter- and intra-class power which gives it sufficient control over the working class to guarantee accumulation, it is also true that workers' struggles repeatedly undermine such control and thus rupture the efficacy (from capital's point of view) of such a class composition. Such a rupture occurs only to the degree that workers are able to *recompose* the structures and distribution of power among themselves in such a way as to achieve a change in their collective relations of power to their class enemy. Thus the struggles which achieve such changes bring about a 'political recomposition' of the class relations – 'recomposition' of the intra-class structures of power and 'political' because that in turn changes the inter-class relations.

In response to such an overcoming of its structure of control, of some particular configurations of its mechanisms of domination, capital (i.e. the managers of production) must seek to 'decompose' the workers' newly constructed relations among themselves and create some new, controllable class composition. The introduction of new technologies, of new organisations of machinery and workers, if successful, results in the undermining of workers' struggles and their reduction, once more, to the status of labour-power. But whatever new 'class composition' is achieved, it only becomes the basis for further conflicts because the class antagonism can only be managed, it cannot be done away with. Thus, these three new concepts, one static and two dynamic, provide guides to the analysis of what have come to be called 'cycles of class struggle', wherein the upswing in such a cycle involves a period of political recomposition by workers and the downswing, however much the workers win or lose, a process of class decomposition through which capital re-establishes sufficient control to continue its overall management of society.[12] The concept of political recomposition theoretically articulates the central role of working-class struggle at the heart of technological change and the concepts of class composition and decomposition provide vehicles for rethinking the issue of technological domination in terms of capital's efforts to cope with an autonomously active, and opposed, historical subject. These concepts both complement and extend Marx's analysis. In Italy they constituted not only a theoretical and political challenge to the hegemony of CPI-style Marxism but more importantly they provided the Italian New Left, and then others elsewhere, with partial guides to a politically useful research agenda geared directly to the development of workers' struggles.

With the development of the class struggles in Italy – especially with the rise of the student and then the women's movement and community conflicts in general – these new concepts were extended from the analysis of the sphere of production narrowly defined to a much broader analysis of the whole of capitalist society. The theoretical basis for such an extension already existed in Mario Tronti's analysis of the tendency of capital to extend its domination from the factory to the rest of society, to transform society into a 'social factory'.[13] If such theoretical considerations had indicated that the 'reserve army' was not really in reserve at all but actively put to work in the circulation and reproduction of capital (and thus part of the working class), the rebellious self-activity of 'unwaged' students and housewives convinced the Italian New Left that they were integral parts of the working class for-itself as well and the analysis of class composition must include the totality of the working class.[14] The political recomposition of the working class was thus understood to involve not only changes in the distribution of power among waged workers in the factory, but also changes among the unwaged and in the relationships between the waged and the unwaged.[15]

With this example in mind, I want to move to the examination of another area of theory where we need, and have recently made strides towards achieving, the kind of theoretico-political inversion necessary for the full development of Marx's theory where he left it lopsidedly underdeveloped. This second area of Marxist theory is the one surrounding the concept of 'valorisation'.

From Valorisation through Alienation to Disvalorisation

Marx's theory of valorisation is at the core of his theory of capitalism. 'Valorisation' *(Verwertung)* designates the complex process through which capital is able not only to put people to work, but to do so in such a way that the process can be repeated on an ever greater scale. Technically, valorisation involves all of the steps included in Marx's circuit of productive capital: the process of production, wherein people are put to work producing products which exceed their own requirements for living, the sale of those products at prices which permit the realisation by the capitalist of surplus value, and finally the reinvestment of that surplus value such that people will, once more, be put to work. To label this process 'valorisation' is to emphasise the enormous transformation capital achieves by reducing the diversity of human productive activity to a unified mechanism of social control. Marx's analysis of this process in terms of 'value' captures the essentially undifferentiated sameness of the production activities included within this process from the point of view of capital. It makes no difference what kind of production is undertaken, what kind of work is done, as long as it produces a product whose sale will realise enough surplus to make it possible to begin all over again.

This undifferentiated sameness can be seen not only in the concept of abstract labour and in the capitalists' indifference to the kind of labour commanded, but also in the money form, which directly symbolises the qualitative equivalence among both products and the labour which produced them, and through profit, which is the money form of command over future labour – money which can be used to renew the original kind of work, or to command some other (if more profitable) form of labour. The labour theory of value is thus a theoretical expression of capital's own view of work and the meaning of work in society. But, unlike other theories, such as neoclassical economics, which in its own way also expresses capital's perspective, the labour theory of value makes the alienating reductionism of capitalist command transparent and provides conceptual tools for a quantitative as well as qualitative understanding of the dynamic of capitalist domination. Thus the theory of value is inextricable from the theory of surplus value, the theory of the labour process from the theory of valorisation, the theory of society from the theory of the accumulation of life as work.

Put differently, the theory of valorisation is the theory of the way in which capital subordinates, transforms and utilises human productive activities for its own purpose: endless command over society.[16] To understand this is to understand why Marxists have sometimes expressed the nature of capitalism by saying that it is a case of 'production for the sake of production', or, to put it differently, capital puts people to work most fundamentally just for the sake of putting them to work – it is capital's way of organising civilisation as one vast labour camp – the global Gulag.[17] This is an understanding which points toward the kind of theoretico-political inversion of class perspective which can enrich our understanding of valorisation and the class struggles associated with it.

Marx and Alienation

Once we recognise that valorisation involves the subordination of human productive activities to capitalist command, we can, as in the case of the 'composition of capital', invert our perspective and examine this phenomenon from the point of view of the people whose productive activities are being subordinated. When we do this we can draw together a variety of moments of analysis from the history of Marxist analysis of work. Qualitatively, Marx addressed this issue directly when he discussed the way in which capitalist command over people's work results in *alienation*. The counterpart of capital's control over the labour process, over the relations among workers and over the product, is the workers' experience of alienation: lacking control, they experience work as in imposed, forced activity rather than as a self-determined activity, they are separated from and pitted against their co-workers rather than finding in work one interesting form of social interaction, and their products are used against them rather than being expressions of their own personalities and vehicles for bonding with others. Thus in his discussion of these phenomena in the *1844 Manuscripts*, Marx dwells both on how the workers are reduced to objects in these processes and the workers' feelings about them.[18] Later on in the *Grundrisse* and *Capital*, while he would emphasise and elaborate on the former, he would take the latter for granted. Even more vividly than in his earlier methodical examination, the capitalist imposition of work appears repeatedly in *Capital* as a 'Werewolf or Vampire-like' relationship through which capital maintains itself and workers experience their subordination as the draining away of their life.[19] Through these texts we can understand Marx's theory of 'alienation' as an inverted perspective on 'valorisation'.

However, the isolation and objectification of the subject denoted by the concept of alienation fails to reflect a fundamental aspect of valorisation which we have examined, namely the reductionism achieved through the processes of subordination. In and of themselves productive activities are enormously diverse and involve many kinds of human activity, yet as they become subordinated to capital not only are they treated on the common ground of being means to social control, but over time, with the development of what Marx called the 'real subordination of labour to capital', the diversity of workers' activity is reduced through a material simplification in which most workers are divorced of their skills, knowledge and mastery of production which are, in turn, concentrated both in the minds of a much smaller number of workers and in the fixed capital of machines. This, of course, is a side of the story of capitalist domination which even Adam Smith recognised (and decried): the degradation of workers from craftspeople to cogs in an industrial machine. Marx's analysis of alienation touched on this, but he dealt with it in much greater detail in his later works, especially the *Grundrisse*, where, in the so-called 'fragment on machines', he projected this tendency to the point where workers are reduced to mere tenders of machines, barely essential to the production process itself. Subsequently, Marxists of various stripes have recognised that technological development, especially but not uniquely in the factory, has involved such degradation of workers' skills. Indeed, another way of talking about this is in terms of 'deskilling'.

Two moments of such deskilling that have received considerable attention were the development of Taylorist and then Fordist methods of reorganising production and workers' tasks. In the case of Taylorism, it has been pointed out how Frederick Taylor was quite self-consciously involved in undermining workers' ability to limit the amount of work they could be forced to do for capital by utilising the power they possessed over the labour process.[20] The stop-watch and clipboard were the tools not only to appropriate the workers' skills but to decompose them (and the workers' power) in such a way that the capitalist rather than the workers controlled the process – and thus the rhythm, continuity and intensity of work. Fordism, in turn, to the degree that it involved a reorganisation of the labour process around the assembly line, also organised a further reduction of worker skill and command over work.[21] More recently, with the growth of the service sector and the extension of Taylorist and Fordist methods to office and other kinds of white-collar work, some Marxists have also extended their analysis of these forms of domination to those new arenas.[22]

Illich and *Desvalor*

Also recently, a growing concern with how the deskilling of many has had as its counterpart the creation of small groups of skilled 'professionals', as well as anger over the displacement of peasant cultures by a spreading capitalist cultural imperialism, has led some non-Marxist intellectuals, who have nevertheless been clearly influenced by Marxism, to elaborate a concept which they call, in Spanish, '*desvalor*', or 'disvalue'. This concept has a theoretical content very close to what one might look for as an inversion of 'valorisation'. The concept of 'disvalue' seems to have originated in the work of Ivan Illich, who, following in the path of Karl Polanyi, has for quite some time been elaborating a critique of 'market society'.[23] A central aspect of Illich's critique, which he spelled out at some length in his book *Tools for Conviviality*, has been the analysis and rejection of both the commodification of needs and the professionalisation of their satisfaction.[24] The emphasis on commodification is very Marxist in two ways. First, in historical terms, Marx's own analysis of capitalism emphasised the tendency of capitalism to take over and commercialise all aspects of life. A central aspect of his analysis of primitive accumulation, for example, was the displacement of domestic food and handicraft production by capitalist commodities – an essential aspect of what he called 'the creation of the home market'.[25] Second, in theoretical terms, the commodity plays a central role in Marx's analysis of capitalism; it is the universal form of wealth, and thus one universal embodiment of value and the class relations of capitalism.[26]

The second part of Illich's critique appears as an updating of Marx's and many Marxists' analysis of the separation of manual and mental labour and its negative impact on workers.[27]

Whereas Marxist preoccupation was primarily with the factory and later on the office, and the way workers were degraded by having their skills stripped from them, and monopolised by higher paid 'mental workers' such as engineers who alone understand of the design of machinery incorporating those skills, Illich elaborated a similar analysis of the professionalisation of the service sector and replacement of the self-production of use-values by the consumption of commodities. For example, Illich blasted the monopolisation of the skills of learning by professional teachers and the monopolisation of the skills of healing by health professionals.[28] Related to Marx's analysis of the alienation of the producer from both the labour process and the product but emphasising the experience of people as consumers being dominated by their alienated products rather than as alienated labourers producing those products, Illich described the growing 'disvalue' of peoples' helplessness and dependency on commodities and professionals, i.e. on market supplied services. In some ways his concept of 'disvalue' expresses the same phenomena neo-classical economics call 'negative externalities' – or the 'disutility' which emerges as byproducts of market production – such as pollution, whose nasty effects have no price and therefore tend to be ignored in a market economy.[29] Yet, as he has elaborated the concept, it has a more significant meaning than this, and one much closer to the Marxian concepts of alienation in production and deskilling.

From *Desvalor* to Disvalorisation

The usefulness of Illich's concept becomes clearer if we shift our attention from his search for phenomena that can be labelled *disvalue* to looking at the *processes* which produce them, i.e. at processes of *disvalorisation*. As a process, disvalorisation can be seen to express precisely the counterpart of valorisation. That is to say, if *valorisation* denotes the capitalist subordination of human productive activities to capitalist command, then *disvalorisation* expresses people's loss of those abilities which are absorbed by capital. Viewed in this way disvalorisation is a more comprehensive and meaningful concept than deskilling or the degradation of work. Although what capital absorbs are carefully and narrowly defined abilities (as in the case of the time-motion studies of Taylor), Illich's treatment shows us that what people lose is much broader; they lose the very fabric of the self-construction of their lives. Those 'abilities' or 'skills' that they lose are integral moments of their own self-determined interconnections with the world, of the sinew of people's lives which give them form and hold them together. In processes of disvalorisation what were integral moments of that sinew are ripped out, isolated and stripped of all their rich interconnections and meanings; they are reduced to narrowly defined skills devoid of their previously rich cultural significations.

Moreover, there is another kind of impoverishment associated with such processes of disvalorisation: namely, the absolute losses which occur when the particularity of diverse skills and abilities are replaced by some narrower range of mechanised, commercialised, mass produced skills. The rise of professional medicine, for example, not only produced a widespread loss of abilities to heal, but it also involved the substitution of one particular paradigm of healing for a much larger number of approaches to 'health', and thus an absolute social loss – the virtual disappearance of a multiplicity of alternative 'values'. If valorisation involves a great reduction of diversity via the imposition of only one relevant measure, then we must recognise that the actual historical processes of disvalorisation are closely connected to *de*valuation or the absolute loss of values.

Marx and Disvalorisation

In Marxist theory *de*valuation has always involved the destruction of capitalist value, e.g. the collapse in existing value that occurs because of a rise in productivity (the falling average value of a commodity reduces the value of existing similar goods) or of an economic crisis (not only the drop in monetary values in times of deflation, but the absolute destruction of unused and wasted resources). If we think of the extreme diversity of abilities and skills which have existed separately outside of capital, irreducible to any common measure, and if we want to speak of those abilities and skills in terms of 'values', then clearly the concepts we use to represent the material diversity must involve an equally great diversity in concepts of 'value'. The counterpart to our refusal of capital's material and conceptual hegemony must not involve the replacement of one hegemony by another but must rather involve the acceptance of non-comparability and of the diversity of 'values'.

It must be said that while Marx was certainly clear about the nature of valorisation and devalorisation within the on-going accumulation of capital, he was ambivalent in his treatment of the processes of devalorisation and disvalorisation during the rise and spread of capitalism. When we examine his writings on primitive accumulation and colonialism – from the *Communist Manifesto* to *Capital* – we often find little or no empathy for the cultures being destroyed/ subsumed by capital. He certainly recognised such destruction/ subsumption but frequently saw its effects on feudalism and other pre-capitalist forms of society as historically progressive. For Marx, workers were being liberated from pre-capitalist forms of exploitation (they 'escaped from the regime of the guilds') and peasants from 'serfdom' and 'the idiocy of rural life'.[30] Yet at the same time, he also showed how they were being hurled into a new form of exploitation and how these changes involved their impoverishment, degradation and enslavement. His treatment of the theft of their independent means of production and living vividly details the violent enclosures and clearings of peasants and artisans from the land, the robbery of their land rights, the destruction of their homes and villages.

Such thoroughgoing devastation meant, of course, the destruction not only of farms, houses and villages, but ways of life, of cultures. But of this we gain little insight from Marx. In his city-boy ignorance of rural life and perhaps in a desire to avoid any backward-looking sentimentalism, Marx seems to have spent little time or energy during his studies of primitive accumulation in England and in the colonies trying to understand what positive values might have been lost. Unlike many of his generation who did worry about the nature of those social ties and communal values which were rapidly disappearing, Marx kept his attention fixed firmly toward the future.[31] This appears to have been first and foremost a political orientation based on his belief in the thoroughness with which capitalism appeared to be quickly wiping out all survivals from its pre-history. If such erasure were being rapidly accomplished, little purpose could be served by dwelling on a defeated past. Only when he thought there was some chance of a pre-capitalist society providing the basis for an effective anti-capitalist struggle did he pay close attention to such phenomena. This exceptional attention we can find in his serious study of the *mir* or peasant commune in Russia. Drawn into a debate over revolutionary strategy in that country, Marx learned Russian, read everything he could find on the subject and concluded that the nature and strength of the *mir* was such as to make it, just possibly, the point of departure for the creation of a communist society.[32]

Time and experience have proven that while the *mir* failed to realise its possibilities, Marx's less informed and much bleaker view of the destiny of other 'pre-capitalist' social forms often both overestimated the efficacy of capital's destructive power and underestimated their tenacity and durability. As a result, any attempt to estimate the possibilities inherent in a vast array of struggles against capitalism in the world today must involve the kind of close and empathetic examination which Marx gave to the Russian peasant commune rather than the kind of cursory, superficial attention he gave to other groups of peasants (see below).

From Survival through Vernacular Subsistence to Self-Valorisation

Recognition of the process of disvalorisation of hitherto autonomous abilities and skills opens a whole new realm of inquiry, namely that of the nature and dynamics of such autonomous abilities themselves, independently of the question of their eventual subordination by capital. Partly, this has been one interest of academic anthropologists who study past, or existing, 'pre-capitalist' cultures, or cultures that have not yet been completely disvalorised into one more impoverished moment of capitalist hegemony. Among political activists in such communities who fight against such absorption – to whom more politically motivated anthropologists sometimes give aid and lend their prestige – many existing non-capitalist cultural practices are not only seen as sources of strength against capital but also as desirable in their own right or as legitimate points of departure for the elaboration of autonomous ways of being.[33] In such circumstances, being able to clearly differentiate between autonomous practices and those which contribute to capitalist valorisation is clearly a necessity for any political strategies geared to the preservation and elaboration of existing cultural autonomy.

On the other hand, however valid such struggles are, this vision is too restrictive. It is too 'historical' in the sense that it is mainly preoccupied with cultural *survivals*, remnants of the past worth preserving and developing into the future. But the genesis of autonomous activities, while inescapable from their historical roots and context, occurs through processes which are constantly renewed. The sources of such autonomy are not merely historical *habits*, which we may associate with daily routines or with periodic rituals, but also include the wellspring of human creativity itself which repeatedly breaks free of habit, whether those habits are the ones cultivated within capitalist valorisation or within some other social framework. Capital itself knows that even where it has completely disvalorised pre-existing abilities and skills, its job is never done.

There is no one term in Marxian theory which expresses capital's own point of view on the activities which constitute such creative autonomy. Such activities are either regarded as creative and imaginative – where they appear to be co-optable – or as deviant and subversive when they resist co-optation and prove to be unrecuperable. The first of these attitudes – apparently open but actually repressive because aimed at co-optation – are important for capital's own development. Indeed, absolutely central to what dynamism exists within capitalism is its ability to absorb, co-opt, or instrumentalise the ever resurgent autonomy of those it has impressed into its 'working' class. The second, more overtly repressive, side is also important for capital because that which it cannot digest it must purge or be poisoned by. Therefore, in mainstream bourgeois social theory there have been many terms to characterise unintegrated, unmanageable working-class autonomy: deviant, delinquent, deficient, uneducated, primitive, backward, underdeveloped, criminal, subversive, schizophrenic, infantile, paranoid, sick, and so on. In the struggle against the repression such terms justify, we must investigate the nature of such autonomy and its relationship to capital's own valorisation with great care. In doing so we should examine any theory – Marxian or non-Marxian – which illuminates the nature and dynamics of such autonomous struggle in ways which help us invert capital's own repressive perspective.

Among non-Marxists, Ivan Illich and his collaborators are among the most interesting of those who have shown a sensitivity to the existence of autonomous creativity in the struggle against contemporary society as well as a desire to contribute to its flowering.[34] Although, like Marx, most of Illich's work has been devoted to detailing the evils of modern society and their disastrous consequences, he has also searched both in the past and in the present to discover alternatives. Early in the 1970s, in the midst of his work on the rise of the service sector (e.g. schooling, the health industry) he emphasised not only the way autonomous competences had historically been converted into needs and associated commodities (i.e. disvalorised) but also the possibilities of developing what he called 'convivial tools' to facilitate either the survival and development of such competences or their genesis.[35] By the early 1980s, Illich's investigation of past forms of autonomy – their nature, their suppression or their survival – led him to shift his emphasis from propositions for a future 'conviviality' to an exploration of concrete 'vernacular subsistence', i.e. autonomous values and practices through which people have satisfied their everyday needs despite and against the depredations of the 'economy'. It is quite clear in his choices of illustrations of his meaning – especially his discussion of language and housework – that this concept is broader than his earlier preoccupation with production and 'convivial tools'.[36] For Illich, the history of the economy (a history – the way he defines it – that most Marxists would equate with that of capitalism) has been a history of a war on autonomous subsistence activities (what we might, at this point, call the history of disvalorisation).[37] There has been such a war because such subsistence activities have both survived and been repeatedly created anew – more so in some places than in others. These activities, Illich and his collaborators feel, provide a point of departure for the elaboration of concrete alternatives to economic society. Gustavo Esteva, for example, who works with Illich and who has been deeply involved with the struggles of urban 'marginals' and rural peasants in Mexico, has eloquently described the nature and struggles of such 'vernacular subsistence activities' to carve out more space for their autonomous development.[38] Along the way, he has contributed to the struggles of pro-peasant *campesinistas* to combat the widespread *assumption* that the peasantry is doomed to extinction. Before confronting such understanding with Marxian theory, let us first examine what there is in the Marxist tradition that recognises and analyses the existence of such autonomous activities, their relationships with capital and their relationship with a possible post-capitalist future.

Marx on the 'Future in the Present'

Marx's own work, from an early stage, emphasised that the germs of the future are to be found in the present. 'Within the old society', he wrote in the *Manifesto*, 'the elements of a new one have been created'. Those elements, of course, were first and foremost to be found in the working class which capital itself had created, but whose autonomous self-activity would lead to the overthrow of the old society and the constitution of a new one. Refusing utopian projects formulated *for* the working class by outsiders, Marx insisted on the need to work within 'the gradual, spontaneous class-organisation of the proletariat' which alone could give rise both to the power to overthrow capitalism and to the power to construct a new social order. The search for the future in the present, he argued, must focus on the struggles themselves.[39]

His own contributions to the identification of such autonomous elements were primarily theoretical. His efforts to locate and understand the forces emerging in opposition to capital and with the potential power to found an alternative to it led him to a focus on the labour process – capital's fundamental vehicle of social control. There, at the heart of capitalist power, Marx isolated and emphasised the autonomous creative moment within it: 'living labour'. Indeed, Marx followed Hegel in seeing in the creativity of living labour an essential aspect of what made humans different from the rest of nature. However anthropocentric this view may have been, when coupled with the perception that the major mechanism of capitalist domination was the control of that living labour, it resulted in Marx's privileging the analysis of the labour process and locating within it one crux both of the class struggle and of the possible transcendence of capitalism. Living labour was at the heart of the class struggle because its dynamism was a fundamental source of antagonism against capitalist domination. He saw the struggle for freedom from domination as being located, in part, in the struggle by creative living labour to liberate itself from outside control. 'The ontology of living labour is an ontology of liberation'.[40] In turn, capital was forced constantly to adapt to that antagonism by seeking to harness the strength of that creativity to its own valorisation. While capital's socialisation of labour would clearly be the point of departure for 'associated labour' in a post-capitalist society, his analyses of the way capitalist control has been embodied in existing forms of 'socialised' labour – such as his theory of the 'composition of capital' – suggests that while current 'socialisation' is relevant to the future, the concrete forms that it presently takes will undergo substantial modification.[41] Apparently, the weakness of workers' struggles in the nineteenth century gave Marx little or no opportunity to study which aspects of socialised labour might be perpetuated by workers in a post-revolutionary period and which aspects might be eliminated or transformed.

At the same time, other parts of Marx's writings made clear that living labour was only part of the source of antagonism opposing capital's domination and fighting to build a new world; there was also the struggle against the reduction of human life to the one-dimensionality of work, and for the creation of time and space for a many sided existence.[42] His most eloquent discussions of the struggle against work can be found in the historical parts of chapter 10 of volume I of *Capital* on the working day where he chronicled workers' battles to reduce work, and in the more abstract 'Fragment on Machines' in the *Grundrisse*, where he evoked the possibilities of liberation from work and the substitution of 'disposable time' for 'labour time' as a measure of value.

Unfortunately, however, Marx's detailed studies of concrete struggles against work were not complemented by anything like an equally detailed study of worker attempts to fill the time liberated by such struggles. This, together with his tendency to dismiss the struggles of pre-capitalist survivals, meant that he left us little in the way of exemplars of such study. He spoke evocatively of the abstract possibilities of workers converting the socialised labour of capitalism into a post-capitalist associated labour, of the liberation of time from work creating the space for community *(Gemeinschaft)* and the 'free development of individualities', but he failed, except in the case of the Russian *mir*, to identify any concrete developments in the present which could be seen as 'elements of the future'.

In his analysis of the Russian peasant commune, Marx's analysis goes a little further than his more general remarks about the potentialities inherent in socialised labour under capitalism. In this case he analysed the 'dualism' of this form of social organisation, isolating both those forces pushing towards its disvalorisation into capitalism and those pushing for its autonomous development. His discussion of those aspects of the peasant commune whose development – because of its nature and omnipresence – could make it the 'fulcrum for social regeneration in Russia' centred on common land ownership, collective cultivation of common meadows, and the *artel* relationship (or traditional practices of cooperation in production or housing construction). All of these practices, he argued, provided concrete moments in the development of social cooperation in labour and appropriation.[43] If a revolution in Russia, he argued, could destroy the state and capitalist exploitation of the peasants and make available to them the material 'positive achievements' of Western capitalism, then the 'spontaneous development' of their communal agrarian organisation could lead directly to a new post-capitalist society.[44]

The two major tendencies of subsequent Marxist theory (orthodox and Western) both ignored much of his analysis. The first tendency was an impoverishing one: it downplayed the struggle against work as 'economistic' in favour of the guiding role of the Leninist party, while embracing a narrow vision of socialism as the liberation *of* work. In its fascination with a one-class society of workers, this productivist tendency conveniently forgot Marx's understanding that living labour develops most fruitfully when it exists as only one moment of a broader, more diverse life experience.[45] The second tendency, which fully understood these limitations in the first, was more depressing than impoverishing: although its studies have followed capital beyond the factory into its efforts to colonise the cultural time liberated by working-class successes in reducing work time, it has simply expanded the orthodox vision of despotism in the factory and catalogued every clever, manipulative mechanism of cultural domination it could find.[46] Neither tendency has ever proved capable of developing a theory of working-class autonomy as an effective force against capitalism and both have always privileged the role of intellectuals (i.e. their own role) as the key to successful social transformation.

Outside these main streams of Marxist tradition, there have been some who remembered Marx's own preoccupation with the power of working-class autonomy which they too sought to understand and to augment. After the cycle of struggles of the late 1910s and early 1920s, the council communists emphasised the creative moment of working-class struggle which had given birth to the soviets in 1905 and 1917 and to the workers' councils in Western Europe after 1918.[47] In the 1950s, a variety of non-Leninist autonomist Marxist groups accorded the same respect not only to the workers' councils created in the upheaval of the Hungarian Revolution of 1956, but also to the day to day ability of workers on the job and in their communities to create new kinds of social relations.[48] This recognition and respect was accorded not only to the autonomy of workers in general but also to various sectors of the class, such as black workers *vis-à-vis* whites and women *vis-à-vis* men.[49] This was a kind of appreciation of diversity which had been rare among Marxists but which would be demanded with great vigour first by the minority movements in the 1960s and then by the women's movement in the 1970s. These Marxists spoke of 'the invading socialist society' and sought, more carefully than Marx himself, to identify those autonomous moments of concrete working-class creativity which might prefigure post-capitalist society. The purpose of such identification, of course, was to found political strategies to strengthen such positive moments of struggle.

Self-Valorisation

Then, out of the international cycle of struggles of the late 1960s and 1970s came a new theoretical articulation designed to express precisely this autonomous creativity as a fundamental source not only of working-class power but of the possibility of going beyond capitalism. As one contribution to a whole movement to build, as well as to theorise, the development of working-class autonomy within and against capitalism, Italian Marxist Antonio Negri proposed the concept of working-class '*autovalorizzazione*' or 'self-valorization'.[50] This concept grew out of the early work by Panzieri, Tronti and others to grasp simultaneously the full extent of capitalist power (such as its attempts to convert all of society into a 'social factory') and the full potential and expression of the working-class power of 'refusal', of its power to subvert capitalist domination.[51] Negri's concept of 'self-valorisation' aimed at contributing to the latter project by showing how the power of refusal could and must be complemented by the power of constitution. In many ways his concept expressed the side of workers' struggles, especially those of young workers, which was coming to the fore in the late 1960s and early 1970s: the creative use of times, spaces and resources liberated from the control of Italian and multinational capital – uses such as the proliferation of 'free radio stations' or the widespread development of women's spaces which, along with many other self-managed projects, helped constitute what many came to call 'the counter-culture'.

Although Marx sometimes used the term 'self-valorisation' as a synonym for 'valorisation'. Negri proposed an entirely distinct meaning. His use of the prefix 'auto', or 'self' (which sounds more natural in English translation), indicates a process of valorisation which is autonomous from capitalist valorisation – a self-defining, self-determining process which goes beyond the mere resistance to capitalist valorisation to a positive project of self-constitution.[52] 'The self-valorisation of the proletarian subject', Negri writes, 'contrarily to capitalist valorisation, takes the form of auto-determination in its development'.[53] Through a close textual reading of the *Grundrisse*, Negri argued that the concept of self-valorisation was implicit in Marx's development of his concept of the working class from labour power through living labour and collective labour to the wage as autonomous power, working class for-itself and the proletariat as revolutionary subject. We can see that it also designates one aspect of those struggles that had come to be analysed in terms of political recomposition.

Negri's concept of self-valorisation thus designates what I find useful to characterise as the *positive* moments of working-class autonomy – where the *negative* moments are made up of workers' resistance to capitalist domination. Alongside the power of refusal or the power to destroy capital's determination, we find in the midst of working-class recomposition the power of creative affirmation, the power to constitute new practices. In some cases, these autonomous projects are built on old bases, inherited and protected cultural practices from the past that have successfully survived capital's attempts at disvalorisation and devalorisation. In other cases, these

projects are newborn, created fully formed out of appropriated
elements which have hitherto been integral parts of capitalist accumulation. In such cases self-valorisation is not only autonomous from
and opposed to valorisation but it can also be the converse of
disvalorisation. It can include processes akin to what the Situationists
used to call '*détournement*' or the diversion of elements of domination into vehicles of liberation.[54]

The relationship between the refusal of capital's determination
and the affirmation of self-valorising activities is an intimate one.
The power of self-valorisation is largely the power to fill the spaces
liberated from capitalist domination with alternative, autonomous
projects. Thus the importance of the refusal of work, highlighted by
Tronti and others in the Italian New Left (as well as by the French
Situationists), is not displaced but appears as the necessary foundation of self-valorisation. 'The refusal of work', Negri writes, 'its
planned organisation by the working and proletarian class, measures
the quantity and quality of the transition, measures . . . the concrete
constituting process [of self-valorisation] determined by the subject'.[55] If capital is successful in converting all of life into work there
is no space or time or energy for self-valorisation. The refusal of
work with its associated seizure of space (e.g. land, buildings) or
time (e.g. weekends, paid vacations, non-work time on the job) or
energy (an entropy raising diversion from work) creates the very
possibility of self-valorisation.[56]

An important part of Negri's elaboration of the concept of self-
valorisation is his recognition that, unlike valorisation and unlike
most socialist visualisations of communism, it does not designate the
self-construction of a unified social project but rather denotes a
'plurality' of instances, a multiplicity of independent undertakings –
not only in the spaces opened within and against capitalism but also
in their full realisation. Communism for Negri, is thus not only a
self-constituting praxis, but it is also the realisation of 'multilaterality'
of the proletarian subject, or, better, of a subject which in its self-
realisation explodes into multiple autonomous subjects. In this way
his concept embraces the kind of intra-class autonomy recognised
and held sacred by autonomist Marxist groups since at least the
1950s.

The concept has also proved flexible enough to be useful for
understanding and appreciating struggles which have often been
considered outside of the working class. These include not only the
struggles of so-called urban 'marginals' which have often been

relegated to the 'lumpenproletariat', but also a wide variety of peasant struggles. Unlike traditional Marxists who have tended to *assume* backwards or petty bourgeois politics among peasants, Marxists equipped with the concept of self-valorisation have been able to perceive and learn from the diversity of peasant projects of communal construction which do not fit those traditional expectations. In this they have rejoined Marx who, through his careful study of the Russian peasant commune, was able largely to shed the anti-peasant bias he had developed in his more limited studies of peasants in Western Europe. They have also joined forces with Esteva and his *campesinistas* who have insisted, against traditional Marxists, not only on the autonomy of such struggles but on the diversity of those efforts.[57]

The concept of self-valorisation thus complements the earlier reconceptualisation of the capitalist tendency to widen its valorisation to all of the 'social factory'. This not only engenders broader refusal, but also a proliferation in the number and diversity of projects of self-valorisation confronting capital in the spaces opened by that refusal. Self-valorisation thus appears to be a concept rich enough to counterpose to valorisation. Where Marx's concept of valorisation draws our attention to the complex sequence of relationships through which capitalism renews itself as a social system of endlessly imposed work, so the concept of self-valorisation draws our attention *through* the complexity of our refusal of valorisation *to* our efforts to elaborate alternative autonomous projects which constitute the only possible source of a self-constituting alternative to capitalism.

Self-Valorisation and Vernacular Subsistence

When we compare this Marxist approach with Illich's concepts of 'conviviality' and 'vernacular subsistence activities' the similarities seem more striking than do the differences. The kinds of existing subsistence and convivial activities, identified by Illich and his collaborators, certainly seem to be embraced by the concept of self-valorisation or the self-constitution of alternative ways of being. Nor is it hard to see how, in Marxist terms, we can understand the 'war against subsistence' as capitalist attempts to disvalorise (or, failing that, to devalorise) just such autonomous self-valorising activities. The problem, which Illich raises by talking about these activities in terms of the Marxian concept of the production of use-values (as opposed to the production of commodities), lies neither with the ideological taint he detects, nor any tendency to confuse use-values with 'unpaid, standardised, formalised activities'.[58] The problem lies rather in the restriction of the concepts of vernacular or self-valorising activities to the sphere of production. Both concepts, as we have seen, are broader than this. Illich, by his choice of illustrations, and Negri, by his emphasis on the refusal of work, have both avoided precisely the capitalist tendency to subordinate everything to work.[59]

Moreover, both concepts are quite self-consciously aimed at movements from the present into the future. There is nothing in Illich's more recent formulations to contradict his earlier aim of expanding the spaces of conviviality through the intentional development and design of 'convivial tools' – new technologies and ways of being that avoid subordination to 'the economy' (i.e. to capitalism or socialism). Negri's critique of traditional Marxist concepts of the 'transition' from capitalism to communism, in which he argues that the only meaningful transition can occur through a development of self-valorising activities which negates capitalist command, makes clear that the concept of self-valorisation designates the existing ground of an emerging post-capitalism.

Also similar in the two approaches is the appreciation for the diversity of such movements regarding the future. Both Illich and Negri quite explicitly want to escape the homogenising measurement and manipulations of capitalist 'Chicago Boys' and socialist 'Commissars'.[60] Against traditional socialist demands to subordinate difference to unity in the struggle against capital and in the construction of a unified post-capitalist order, both embrace what Negri calls the 'multilaterality' of self-determination, the multiplicity of autonomous projects whose elaboration can constitute a new world whose 'pluralism' would be real rather than illusory as is the case today in the world of capital.[61]

The major differences between the two approaches derive from their conceptualisation of the social setting of 'vernacular' or 'self-valorising' activities. Illich's understanding of that setting – largely derived from Polanyi and Dumont – as an 'economic' society which seeks to reduce human beings to special cases of *homoeconomicus* is close to, but not the same as, a Marxist understanding of capitalism as a social system which seeks to reduce everyone to mere worker. Methodologically speaking, the most important consequence of this difference is that whereas Illich can identify and examine the development of 'vernacular' human activities which have escaped integration into the economic, Negri's Marxist analysis can also grasp such 'self-valorising' activities as both generated by and yet, in their antagonism, autonomous from the dialectic of capital. Illich and those who utilise his approach can observe, and herald, the ultimate failure of the 'economy' to achieve complete hegemony while celebrating the autonomous, convivial subjects whose resistance they accord responsibility for that failure. But, unlike the concept of self-valorisation, that of the vernacular provides no theory of *the genesis* of such antagonistic subjects within the development of contemporary capitalist society. Illich laments the destruction of the subject and calls for the pursuit of 'conviviality' but, because of his rejection of Marxism, fails to show how such autonomous subjects repeatedly arise within even the thoroughly woven nets of capitalist control. This is why, I think, there is so much emphasis among Illich's collaborators on those living in communities that have been able to avoid, more or less, being fully integrated into capital.[62] In particular it helps explain both the preoccupation with the period before and during the rise of capitalism (of Polanyi's disembedded and hegemonic 'economy') and the attractiveness of Illich's theory to those in the Third World (or those who study it) – where perhaps most communities which are readily identifiable as autonomous are located. While there can be no doubt that such communities can certainly be considered privileged zones of insight because of the degree to which they have achieved the elaboration of their own autonomous ways of being, the overthrow of the existing capitalist (economic) order can derive only from the proliferation of a diversity of such spaces throughout the social fabric of capital. We need both an awareness and appreciation of such possibilities – which either approach can provide – and a theory of the processes of spontaneous antagonistic generation which give them birth – which only Marxism has been able to provide.

Beyond Marxism?

But if Marxism provides a theory of the generation of autonomous subjects within capitalism, its relevance to the internal logic of the autonomous development which is constructed within those spaces, to our understanding of and contribution to the processes of self-determination, is not so obvious. If one is a dialectical materialist, of course, or even a historical materialist, for whom Marxism is a philosophy of universal applicability, then Marxism remains relevant, at least methodologically. But if one takes Marxism as a theoretical and political practice which emerged within capitalism and whose usefulness is restricted to a working-class articulation of the class struggles of that society, then understanding the many processes of self-valorisation or self-constitution that escape the control of capital clearly requires grasping those processes in their own terms.

We may proceed analogically at first, seeking to utilise what we have learned by studying capitalism as a point of departure for grasping such uprisings against it, but that will inevitably focus our attention mostly on the scars of the birth process – the ways these self-activities have been shaped by the fact of their emerging out of capitalism. These scars we can see all around us for we inscribe our struggle first on the walls with which capital imprisons us. We liberate spaces and times but those spaces and times are still bounded by the structures of capitalist power. We craft autonomous environments and activities but we do so in spaces scarred by capitalist exploitation and with commodities and personalities at least partially shaped by the processes of valorisation. All this guarantees that post-capitalist society will no more establish itself in the world completely freed of its past than capitalism did. Just as capitalism, even today, is forced to deal with pre-capitalist 'survivals', so too can we know that our efforts to create a post-capitalist society will be marked with the signs of capitalism – for a long time to come. To the degree that this is so, the Marxist theory of capitalism will be of continuing interest.

It is in identifying and understanding the new and original qualities of self-valorising activities that we face the greatest need for imagination and creativity and can rely the least on old theories, even those of Marx. On the one hand, we can expect new understanding to be generated within and as an integral part of such new activities. On the other, if we accept the idea that post-capitalist society is coming into being as a plurality, as a complex mixture of diverse ways of being, then it is already obvious that we, as individuals, are unable to participate authentically in more than a fraction of such alternatives and, as a result, are faced with the task of trying to understand those projects of self-valorisation that we do not know from within. Unlike the theoreticians of capital who can simply project their own concepts on to others as part of the capitalist project of subsumption of everyone and every social structure into itself, the struggle against all forms of domination requires the refusal of such theoretical imperialism and much more open, imaginative attempts to under-stand alternative ways of being in their own terms. It is for this reason that we must privilege all such projects of self-valorisation – those we are involved in and those we only observe from the outside. Even where capital is successful in crushing the autonomy of self-valorisation and subordinating that creativity to itself, the experience or study of that autonomy can inform other efforts to build the future within the present. It has always been true in the class struggle that we need to learn from past mistakes. It has become increasingly true that we also need to recognise and learn from our successes, however fleeting, in constructing autonomous ways of being.

Notes

1. This working-class approach to bourgeois theory is quite different from that practised by most Marxists. Traditionally Marxists either have dismissed mainstream theory as purely apologetic or have criticised various moments of that theory, pointing out their internal inconsisten-cies or their failure to grasp some essential aspect of the world. Ironically, such well intended approaches have had two unfortunate side effects. First, by inducing contempt for mainstream theory the attention of those struggling against capitalism is diverted away from the serious study of that theory as a key to the capitalist strategies being wielded against them. Second, by identifying logical lapses and critical oversights in the theory, such critiques can only help mainstream theorists improve the rigour and usefulness of their formulations – to the detriment of those in whose interests the critics seek to act!

2. Fortunately, a variety of Marxist historians have made major contribu-tions to filling in the missing side to this story, but we have yet to see any systematic attempt to explore the other side of primitive accumula-tion theoretically. For a beginning at such attempt see the discussion of disvalorisation below.

3. This is the case even though the full discussion of surplus value as 'profit' was relegated to the first part of volume III of *Capital*. The account in volume 1 traces not only the actions of the capitalists but also their arguments (e.g. Senior's about the dangers to business profits of any shortening of the working day).

4. Although Marx discusses these concepts at many different points, his clearest exposition of the differences and relations among them can be found in the opening pages of chapter 25 of volume 1 of *Capital* and in volume III of that same work.

5. This is the proper counterpart only in the abstract. All empirical efforts

actually to measure the capital/labour ratio involve methods of aggregation – usually at market prices – which make this concept more akin to the value composition of capital discussed in the next paragraph.

6. Note: the value of the constant capital introduced is irrelevant to the issue of a rise in the organic composition of capital. The only thing that matters is whether the introduction of that new constant capital raises the productivity of labour. Thus the error of those, such as Paul Sweezy, who have argued that there is no inevitable long-term 'tendency of the organic composition of capital to rise' because the per unit value of constant capital falls with the rise in the productivity of Department I industries producing the means of production. See Sweezy's discussion in his *The Theory of Capitalist Development*, (New York, 1942), pp. 103–4, and the subsequent debate.

7. See Raniero Panzieri, 'The Capitalist Use of Machinery: Marx Versus the Objectivists', in Phil Slater (ed), *Outlines of a Critique of Technology*, (Atlantic Highlands, 1980), pp. 39–68. Originally published as 'sull'uso capitalistico delle macchine nel neocapitalismo', *Quaderni Rossi*, no. 1 (1961).

8. Among the most important of these studies were those by Romano Alquati. See, for example, his 'Composizione organica del capitale e forza-lavoro alla Olivetti', (1962) and 'Ricerca sulla structtura interna delle classe operaia' (1965), reprinted in Romano Alquati, *Sulla FIAT e Altri Scritti* (Milan; 1975)

9. As in the passage from chapter 15, section 5, of volume I of *Capital*: 'It would be possible to write a whole history of the inventions made since 1830 for the sole purpose of providing capital with weapons against working-class revolt'.

10. For more discussion of these concepts and their usefulness in analysing the history of working-class struggle, see Harry Cleaver, *Reading Capital Politically* (Austin; Tex., 1979), and Yann Moulier, 'L'Operaisme italien: organisation/representation/ideologie ou la composition de classe revisité', in Marie-Blanche Tahon and Andre Corten (eds), *L'Italie: Le philosophe at le gendarme, Actes du Coloque de Montreal*, (Montreal; 1986).

11. It also opened the door to a reinterpretation of the meaning of the 'relations of production' and 'forces of production' – those concepts whose use in the 'Preface' to the *Contribution to the Critique of Political Economy* has formed the basis of so much unproductive speculation. For such a reinterpretation see Harry Cleaver and Peter Bell, 'Marx's Crisis Theory as a Theory of Class Struggle', *Research in Political Economy*, V.5 (1982), pp. 194–5.

12. There is no assumption here that class confrontation will always assume a 'cyclical' form – that depends entirely on whether capital is actually able to re-establish control. Marxist analysis holds out the constant

possibility that such confrontation may achieve revolutionary success, such that capital fails to re-establish control and is driven from the historical stage.

13. In an essay written in 1962 on 'The Factory and Society', Tronti wrote: 'The more capitalist development advances, that is to say the more the production or relative surplus value penetrates everywhere, the more the circuit production – distribution – exchange – consumption inevitably develops; that is to say that the relationship between capitalist production and bourgeois society, between the factory and society, between society and the state, become more and more organic. At the highest level of capitalist development social relations become moments of the relations of production, and the whole society becomes an articulation of production. In short, all of society lives as a function of the factory and the factory extends its exclusive domination over all of society' (my translation). Mario Tronti, 'La fabbrica e la società', *Quaderni Rossi* no. 2 (1962), p. 20. (This essay was reprinted in Mario Tronti, *Operai e capitale* (Turin 1966 (1971), pp. 39–59).) This theorisation grounded an alternative approach to Gramsci and the Frankfurt School's visions of capitalist hegemony – one in which class antagonism does not disappear but permeates everything, including the dynamics of cultural development.

14. Early analysis of the 'unwaged' as integral parts of the working class was developed in Italy. See, for example, Mariarosa Dalla Costa and Selma James, 'The Power of Women and the Subversion of the Community', *Radical America*, vol. 6, no. 1 (January-February 1972), or Collettivo Internazionale Femminista, *Le Operaie della Casa* (Venice 1975). The subordination of unwaged work – such as housework – to the reproduction of capital was later analysed in non-Marxian terms by Ivan Illich in terms of 'shadow work'. See his book *Shadow Work* (Boston, Mass., 1981).

15. For an example of the extension of these concepts in Italy see Roberta Tomassini, *Studenti e Composizione di Classe* (Milan, 1977). For examples of such analysis – at the level of the 'social factory' – in the United States see Paolo Carpignano, 'U.S. Class Composition in the Sixties', the other articles in *Zerowork*, no. 1 (1975), and no. 2 (1977), and more recent articles in *Midnight Notes*. For an example of such analysis applied to unwaged peasants in the Third World see Ann Lucas de Rouffignac, *The Contemporary Peasantry in Mexico*, (New York, 1985).

16. In short, the theory of valorisation is another part of the Marxist theory of capitalist domination, complementary to the theory of the composition of capital but different from it, for we can discuss valorisation in abstraction from the particular relationship between constant and variable capital. The concept of the composition of capital is at once more

concrete, because it does deal with such particular relations, and more limited, because it always refers to production, while the circuit of valorisation includes not only production but also circulation.

17. The juxtaposition of the 'labour process' with 'valorisation' in chapter 7 of volume I of *Capital* does not make this immediately apparent, but section 2 of chapter 10 of that same volume does, when Marx points out that, unlike previous societies where rulers have imposed surplus labour on others to benefit themselves, in capitalism the imposition of work is endless and independent of the production of any particular use-values, including the luxuries consumed by the capitalist class.

18. See especially the section on 'estranged labour'.

19. The persistence of such vivid representations of alienation in *Capital* gives the lie to those, such as Althusser and his followers, who have sought to disassociate the 'science' of Marx's mature work from its Hegelian, immature predecessors. The workers' product owned and controlled by capital becomes a *'monster'* (*Capital*, New York, Vintage edition, p. 302). It is in chapter 10 that we find this monster pictured by Marx as a *Vampire*: 'Capital is dead labour, which, vampire-like, lives only by sucking living labour, and lives the more, the more labour it sucks', (ibid., p. 342). But dead labour is precisely the products produced by workers, products that have become alien objects, part of capital, which are used to dominate workers. The expression 'sucking living labour' clearly means forcing humans to work, and the more they are forced to work, the more products are produced, the more surplus value is extracted and the more capital thrives. Still further (ibid., p. 353) he speaks of capital's *'were-wolf . . . hunger for surplus labour'* (my emphasis), again the alien monster seeking ever to impose more work. And then in chapter 11 we find in somewhat less colourful language: 'It is no longer the worker who employs the means of production, but the means of production which employ the worker. Instead of being consumed by him as material elements of his productive activity, they consume him as the ferment necessary to their own life-process, and the life-process consists solely in its own motion as self-valorising value'. (ibid., p. 425). Finally, in the chapters on machinery and modern industry there is a whole discussion on how under capital the worker comes to serve the machine rather than vice versa.

20. See Mike Davis, 'The Stop Watch and the Wooden Shoe: Scientific Management and the Industrial Workers of the World', *Radical America,* vol. 8, no. 6 (January–February 1975).

21. See Benjamin Coriat, *L'Atelier et le chronomètre*, (Paris, 1979). Both Taylorism and Fordism concerned issues other than deskilling. Both involved new ways of manipulating wages and work incentives as well as direct control over the production process.

22. See, for example, Harry Braverman, *Labor and Monopoly Capital: The*

Degradation of Work in the Twentieth Century, (New York, 1974) and all the spin-offs from it.

23. Illich's way of characterising the society he wishes to overcome has evolved over time. In the early 1970s, in his *Tools for Conviviality* (New York, 1973), influenced by then current discussions of 'limits' and by his readings of Marx, he referred to it as 'industrialism' or 'industrial mode of production', a term he clearly felt to be more inclusive than 'capitalism' and capable of referring to the social structure of 'socialist' as well as those of Western capitalist countries. Later, in his *Shadow Work* drawing on Leiss, Polanyi and Dumont, he called it a commodity-intensive society'. More recently in their work on the 'archaeology of the modern mind', he and his co-workers have often just called it 'economic society'.

24. Illich, *Tools for Conviviality*. See also his 'Useful Unemployment and its Professional Enemies', written as a postface to the earlier book, in Ivan Illich, *Toward a History of Needs*, (Berkeley 1977) as well as his *Deschooling Society* (New York, 1970).

25. See Karl Marx, *Capital*, vol I, ch. 30: 'Impact of the Agricultural Revolution on Industry. The creation of a home market for Industrial Capital'.

26. Ibid., ch. 1.

27. See, for example, Alfred Sohn-Rethel, *Intellectual and Manual Labour: A Critique of Epistemology*, (Atlantic Highlands, 1983).

28. See his *Deschooling Society*, his *Medical Nemesis: The Expropriation of Health* (New York, 1976) and his *Disabling Professions*, (London, 1977).

29. At some points Illich seems to use these terms interchangeably, as in 'This is a form of *disvalue* necessarily associated with the proliferation of commodities. This rising *disutility* of industrial mass products' (my emphasis) in *Toward a History of Needs,* p. 11.

30. The reference to the escape from the guilds and from serfdom is from *Capital* vol. 1, ch. 26. The reference to being saved from the 'idiocy of rural life' is from the *Communist Manifesto*.

31. For an overview of the nineteenth-century preoccupation with the displacement of community by an atomised capitalist society, see John P. Farrell, 'Reading the Text of Community in Wuthering Heights', *ELH*, no. 56 (1989). This article provides useful references to the nineteenth-century discussions in both literature and social thought.

32. See Teodor Shanin (ed), *Late Marx and the Russian Road: Marxism and 'The Peripheries of Capitalism'* (New York, 1983). As the materials in this collection make clear, this political question led Marx to reconsider the history of primitive rural structures in Western Europe as well. Although he recognised that some primitive agrarian communes in Germany (near his hometown of Trier) had demonstrated enough 'natural vitality' to survive into the nineteenth-century, he apparently

thought them to be isolated curiosities and not worthy of the attention he accorded the much more widespread Russian *mir* (p. 107).

33. Among the many such political collaborations between anthropologists and the struggles of indigenous peoples for autonomy, see the work of those associated with the organisation Cultural Survival and its journal *Cultural Survival Quarterly*.

34. Much broader than any one theory or group of intellectuals have been the social movements which have contributed to the construction of, and reflection on, autonomous social projects. Probably the most important of these, in so far as they have truly sought to develop – in both theory and practice – alternative ways of being have been the gender movement (both women and gays), and the green/environmental movement. Here I limit my discussion to Illich, who has sought to draw general conclusions from a wide variety of struggles, but further work needs to be done examining the practices and thinking which has been generated within these social movements, whether those activities are self-consciously 'Marxist' or not.

35. Illich, *Tools for Conviviality*.

36. Idem, *Shadow Work*, ch. II: 'Vernacular Values'.

37. Ibid., ch. III: 'The War Against Subsistence'.

38. Gustavo Esteva, 'Los Tradifas' O el Fin de la Marginación', *El Trimestre Economico*, vol. L(2), no. 198 April – June 1983), pp. 733–69; 'Para Ser Como La Sombra de un Arbol', *El Gallo Illustrado*, no. 1247 (18 May 1986), p. 17; 'En la senda de Juan Chiles', *El Gallo Illustrado*, no. 1250 (8 June 1986); 'Cocinar la Autonomía', *El Gallo Illustrado*, no. 1276 (7 December 1986), pp. 8–9; 'Las naciones Indias en la Nacion mexicana', *El Gallo Illustrado*, no. 1308 (19 June 1987), pp. 8–10; 'Regenerating People's space', *Alternatives*, XIL, (1987), pp. 125–52; 'Food Reliance and Peasant Self-Management: Bases for the Agrarian Transformation of Mexico', typescript; 'Celebration of Common Men', typescript.

39. See the analysis of 'Critical-Utopian Socialism and Communism', in chapter III of the *Manifesto of the Communist Party*.

40. This is Antonio Negri's formulation. See Thesis 13 in his contribution to this volume.

41. Thus the mistake of those such as Lenin who thought capitalist technologies – such as the Taylorist organisation of production – should be taken over and used by workers in a revolutionary society. See the section on 'Raising the Productivity of Labour' in Lenin's *Immediate Tasks of the Soviet Government* (1918), in V.I. Lenin, *Collected Works* (Moscow), vol. 27. Illich is quite correct to attack such positions among orthodox Marxists in his *Tools for Conviviality*, although he seems to have been unaware of similar critiques from within Marxism, e.g. those

of Western Marxism and of the autonomist Marxists cited in the preceeding discussion of 'class composition'.

42. Among those Marxists who have most clearly articulated this aspect of Marx's thought are Herbert Marcuse, especially in his *Eros and Civilisation* (Boston, 1955), the French Situationists, many in the Italian New Left – such as Mario Tronti with his 'strategy of refusal' – and the American authors of *Zerowork*, a journal which was published briefly in the 1970s, and *Midnight Notes*, which is still published out of Boston. Among non-Marxists whose study of the world led them to a similar understanding, see Bertrand Russell and his elegant essay 'In praise of Leisure' in Vernon Richards (ed), *Why Work? Arguments for the Leisure Society* (London, 1983), pp. 25–34, and Jacque Ellul, 'From the Bible to a History of Non-Work', *Cross Currents*, vol. 35, no. 1 (Spring 1985), pp. 43–8.

43. Also of interest here, in the light of his remarks in the *Grundrisse* about the liberation of life time allowing for the 'full development of individualities' are Marx's comments (accurate or not) about how in the Russian commune, as opposed to more primitive kinship based social forms, the existence of 'the house and yard as an individual family preserve' helped foster 'individuality'. Shanin (ed) *Late Marx and the Russian Road*, p. 120.

44. Marx's analysis of these elements is to be found primarily in a series of letters and drafts of letters written as interventions in the debate in Russia over the role of the peasant commune in revolutionary strategy for that country. These materials are now available in Karl Marx and Frederick Engels, *Collected Works*, (New York, 1989), vol. 24, pp. 346–71. Shanin (ed), *Late Marx and the Russian Road* presents these materials along with several interpretive essays and supplementary materials, including translations of several Russians whose work influenced Marx.

45. This tendency includes virtually the totality of the Marxist-Leninist tradition with its socialist work ethic that mirrors in a secular fashion all the narrowness of its Calvinist counterpart.

46. This tendency includes much of 'Critical Theory' and its off-shoots which have proven incapable of either seeing or theorising working-class struggle except through the perspective of capital's instrumentalisation. Marcuse's work, of course, whatever its limitations, was a notable exception within this tradition because he sought to understand the autonomous dynamism of the forces capital had to control in order to survive.

47. For an overview of the council communists see Mark Shipway, 'Council Communism', in Maximilien Rubel and John Crump (eds), *Non-Market Socialism in the Nineteenth and Twentieth Centuries,* (New York, 1987) or Peter Rachleff, *Marxism and Council Communism* (New York, 1976).

48. These non-Leninist Marxists included those who had been part of the Johnson-Forest Tendency in the US, especially C.L.R. James and Raya Dunayevskaya, and those associated with the early years of the journal *Socialisme ou Barbarie* in France. See C.L.R. James, Grace C. Lee and Pierre Chaulieu, *Facing Reality: The New Society . . . Where to Look for It, How to Bring it Closer,* (Detroit, 1974 (1958)), esp. chapter 1; Raya Dunayevskaya, *Marxism and Freedom* (London, 1975 (1958)) and the collection of *Socialisme ou Barbarie.*

49. On the issue of blck autonomy see, for example, C.L.R. James, 'The Revolutionary Answer to the Negro Problem in the USA' (1938), reprinted in C.L.R. James, *The Future in the Present, Selected Writings Vol. I* (London, 1977). On the issue of women's autonomy see the early essay of Selma James, 'The Power of Women' included in Mariarosa Dalla Costa and Selma James, *The Power of Women and the Subversion of the Community* (Bristol, 1972).

50. The fullest discussion of this concept by Negri available in English is to be found in his École Normale lectures on the *Grundrisse* published as Antonio Negri, *Marx Beyond Marx* (South Hadley, Mass.,) A fresh translation is available from the US publisher Automedia.

51. On the power of working-class refusal, see Tronti, *Operai e capitale.*

52. It is important to note that the prefix 'self', as used here, has no necessary connotations of the individual self but may refer either to the individual or to a complex but collective class subject. Negri, in his own work, has tended to use the term self-valorisation in discussing the macro class subject. The concept, however, can also be useful in thinking about the dynamics of individual autonomy – the kind of micro or molecular struggles addressed by Félix Guattari in his *La Révolution moléculaire* (Paris, 1977, English translation – *Molecular Revolution,* trans. by R. Sheed (Harmondsworth, 1984)) or in his and Gilles Deleuze's two books *Anti-Oedipe: Capitalisme et Schizophrénie* (Paris, 1972) and *Mille Plateaux: Capitalisme et Schizophrénie* (Paris, 1980). Recently Negri and Guattari collaborated to write *Les Nouveau Espaces de Liberté* (Paris, 1985). In Negri's current work 'constitution' has largely replaced 'self-valorisation'. See his chapter in this collection.

53. Negri, *Marx Beyond Marx,* p. 162.

54. The Situationists' concept of '*détournement*' was one of the very few earlier moments of Marxist theory which sought to grasp how the mechanisms of domination could be subverted and used by workers for their own purposes. The dominant Marxist paradigm for thinking about the mechanisms of domination seems to have been derived from Lenin's comments on the capitalist state – they could not be used but must be smashed. Among the Italian New Left theorists, Negri's general concept of self-valorisation was predated by a new understanding of the wage as an expression of working-class power. This too reversed the usual

Marxist understanding of the wage solely as a means of exploitation. Among the most interesting work on the wage as source of working-class power has been that by women Marxists involved in the Wages for Housework Movement who developed an analysis of the role of the unwaged within the overall class composition and a political campaign based on that analysis. See Dalla Costa and James, *The Power of Women*, and Silva Federici, 'Wages Against Housework', (1975) in Ellen Malos (ed), *The Politics of Housework*, (London, 1980).

55. Negri, *Marx Beyond Marx*, p. 166.

56. At the same time, without self-valorisation, the refusal of work merely creates empty spaces susceptible to capital's recolonisation. But this is largely an abstract possibility because, as a rule, the struggle against work is not aimed at replacing work with a vacuous do-nothing 'leisure' – as its detractors often cynically insist – but rather with creating the time and space for all the things people would like to do beyond and despite of their work. In those times and places where the working class has been forced to work so long that it has sought 'free time' purely for rest, that rest, however necessary and understandable, can hardly be seen as anything other than the simple reproduction of labour power.

57. See, for example, Ann Lucas de Rouffignac's sympathetic treatment of the *campesinista* position in her *Contemporary Peasantry in Mexico: A Class Analysis*, (New York, 1985), ch. 2 'The Debate in Mexico Over the Peasantry and Capitalism', or her treatment in 'El Debate Sobre los Campesinos y el Capitalismo en México', *Comercio Exterior*, vol. 32, no. 4. (April 1982), pp. 371–83.

58. See *Shadow Work*, p. 58.

59. Illich's earlier work, such as *Tools for Conviviality*, was not so free from the objection he raised eight years later. Despite arguing that his notion of 'tools' was broader than the usual meaning, his choice of illustrations and his pervasive focus on the sphere of production – including the conversion of reproductive activities to production in the service sector – seemed to retain a preoccupation with work as the one activity which could, at least potentially, give meaning to human life. Indeed, in his juxtaposition of labour to work (chapter 2) he reproduced Engels's distinction between undesirable, nasty labour under capitalism and desirable, free work in post-capitalist society. Such a position not only fails to see or appreciate the great diversity of ways in which human life can be realised but also fails to recognise that the only way work can become an interesting mode of human self-realisation is through its subordination to the rest of life, the exact opposite of capitalism.

60. Illich, *Shadow Work*, p. 58; Negri, *Marx Beyond Marx*, Lesson Eight: 'Communism and Transition'.

61. 'Each step toward communism is a moment of extension and of expansion of the whole wealth of differences. . . . The communist

transition follows at this stage the path which leads from auto-valorisation to auto-determination, to an even greater and more total independence of the proletarian subject, to the multilaterality of its way.' Negri,
Marx Beyond Marx, pp. 167–8.
62. See, for example, the work of Esteva mentioned above. He can celebrate
the historical appearance of what he calls 'common man' (as opposed to
homoeconomicus or traditional man etc.) but he cannot explain his
appearance within capitalism. Esteva's conception that his autonomous
'common man' was 'born in the interstices of society' reflects both his
own preoccupation with the 'margins' and his lack of a theory as to how
such autonomous subjects can emerge within and against the dynamic
of capitalist society. In Marxist terms his self-valorising subjects are not
once thoroughly integrated workers (*homoeconomicus* in the language
of Illich and Esteva) who in their struggle against capital have carved
out time and space for the elaboration of their autonomy to the point of
revolutionary rupture. They are rather those whom capital has failed to
integrate into its expanded reproduction – those whom capital defines
as marginal to itself.

5

Crisis, Fetishism, Class Composition

JOHN HOLLOWAY

The world is changing rapidly. Change is often the subject of conflict. Those opposed to change are often depicted as being unreasonable, as foolishly standing in the way of the inevitable, as waging the struggles of a bygone age. Social trends are said to be inescapable.

These arguments have been heard frequently in the last few years, not only on the right, but on the left as well. It is widely argued that capitalism is entering a new stage, often called neo- or post-Fordism, and that socialists must adjust to this new reality and rethink the meaning of socialism.

But is capitalism entering a new phase? And, if so, how did it get there? Is one phase, one inescapable tendency, simply replacing another? If not, what is the nature of the transition? The question is important, theoretically and politically.

The very concept of a 'phase' of capitalism suggests that there is a qualitative turning point, a break in the normal process of change. Social change, always present, intensifies in a way that makes the result of the change qualitatively distinct from the period that went before.

Crisis

A qualitative turning point, a break in the normal process of change, is a crisis. The origin of the term 'crisis' is medical. In its original Greek meaning it referred to a turning point in an illness 'when death or recovery hangs in the balance' (Rader 1979, p. 187, cited by O'Connor 1987, p. 55). Crisis was said to occur in a disease

145

'whenever the disease increases in intensity or goes away or changes into another disease or ends altogether' (Stern 1970, cited by O'Connor 1987, p. 55). In a medical sense, then, a crisis is not necessarily a bad thing. It points rather to the unevenness in the progression of an illness, to the punctuation of relatively homogenous patterns of development by moments in which change for better or for worse is intensified, in which one pattern of development is broken and another (perhaps) established: a time of anxiety and a time of hope.

Applying the concept to social and historical development, crisis does not simply refer to 'hard times', but to turning points. It directs attention to the discontinuities of history, to breaks in the path of development, ruptures in a pattern of movement, variations in the intensity of time. The concept of crisis implies that history is not smooth or predictable, but full of shifts in direction and periods of intensified change.

If history is not a smooth, even process of development, then it follows that the concept of crisis must be at the centre of any theory of social change. As O'Connor (1979) puts it, 'the idea of crisis is at the heart of all serious discussions of the modern world' (p. 49).

Periods of intensified social change can be seen from two angles: as periods of social restructuring, periods in which the social relations of capitalism are reorganised and established on a new basis; or as periods of rupture, of potential breakdown, periods in which capitalism comes up against its limits. As the medical analogy implies, the patient may recover, but may not. The doctor looks at the crisis and examines it for recovery; the grave-digger looks at the crisis with something quite different in mind. The crisis of capitalism has a very special significance for those who would be capitalism's grave-diggers. For the indifferent observer, crisis is a period of intensified change which may lead one way or another; for the person who wants a radically different future, it is the element of rupture which attracts attention.

If the concept of crisis is important for any theory of social change, it is absolutely crucial for any theory that looks at capitalism with an eye to its radical transformation. This is particularly true for the Marxist tradition. What distinguishes Marxism most obviously from other forms of radical thought is the idea that an understanding of capitalism can show not only the *desirability* or *necessity* of establishing a different form of social organisation, but also the *possibility* of doing so. The radical transformation of society is possible because capitalism is inherently unstable, and this instability is expressed in its periodic crises, in which capitalism is confronted with its own mortality. The concept of crisis is at the core of Marxism. It is not an exaggeration to say that Marxism *is* a theory of crisis, a theory of structural social instability. Whereas other radical traditions focus on the oppressive nature of capitalist society, the distinguishing feature of Marxism is that it is not just a theory of oppression but also, and above all, a theory of social instability.

If Marxism is a theory of crisis, it is an open theory. Marx himself did not leave any fully worked out theory of crisis, and the debates on the theory of crisis have continued ever since *Capital* was published. Within the Marxist tradition, there are major differences on crisis theory, between disproportionality theory, underconsumptionist theory, overaccumulation theory, etc. These debates are often conducted in what appear to be technical, economic terms: what is at issue, however, in any discussion of crisis is the understanding of capitalist instability and the possibility of a transition to a radically different type of society. The theory of crisis cannot be separated from our understanding of capitalist society and of what makes it change.

Marx's Concept of Social Change

Capitalism is unstable because it is an antagonistic society. It is social antagonism that is the source of change in society. As Marx put it in the famous opening sentence of the *Communist Manifesto* 'The history of all hitherto existing society is the history of class struggle'.

Within the Marxist tradition there are, however, different ways of conceptualising social change. Sometimes the differences are presented in terms of a distinction made between the young Marx and the old Marx. According to this view, the young Marx stressed struggle and subjective action as the source of historical change, while the more mature Marx, the Marx of *Capital*, analysed social development in terms of the 'objective laws of capitalist development'. In recent years, this distinction has been made most starkly by Althusser and the structuralist school of Marxism, but the (explicit or implicit) separation of struggle from the laws of capitalist development is widespread within the Marxist tradition. Very often, the importance of class struggle is recognised, but it is seen as subsidiary to, or as taking place within the framework of, the laws of capitalist development.

The different emphases can be found not only in differences between the 'young Marx' and the 'mature Marx', but throughout his work. The text long regarded as the classic statement of Marx's theory is the passage in the 1859 Preface to his *Contribution to the Critique of Political Economy* in which Marx presents the conclusions of his early studies:

'The general conclusion at which I arrived and which, once reached, became the guiding principle of my studies can be summarised as follows. In the social production of their existence, men inevitably enter into definite relations, which are independent of their will, namely relations of production appropriate to a given stage of development of their material forces of production. The totality of these relations of production constitutes the economic structure of society, the real foundation, on which arises a legal and political superstructure and to which correspond definite forms of social consciousness. The mode of production of material life conditions the general process of social, political and intellectual life. It is not the consciousness of men that determines their existence, but their social existence that determines their consciousness. At a certain stage of their development, the material productive forces of society come into conflict with the existing relations of production or – this merely expresses the same thing in legal terms – with the property relations within the framework of which they have operated hitherto. From forms of development of the productive forces these relations turn into their fetters. Then begins an era of social revolution. The changes in the economic foundation lead sooner or later to the transformation of the whole immense superstructure'.

(1971, pp. 20–1).

The 1859 Preface has been much criticised in recent years: this has been part of the more general criticism of Communist Party 'orthodoxy' and of the changes in the international Communist Party movement since the 1960s. Usually, the criticism emphasises the 'relative autonomy' of the superstructure, arguing that the economic is determinant only 'in the last instance'. There is thus more scope for achieving social change through the political, ideological or legal spheres than the 1859 Preface would appear to allow.

This criticism appears to make a radical break with the economic determinism of Marx's text. On reflection, however, it can be seen that this argument actually reproduces the same conceptual framework as the 1859 Preface. Society is still analysed in terms of structures, economic, political and ideological; the difference lies only in the autonomy attributed to each of these structures.

There is a more fundamental criticism of the 1859 Preface – which indeed can be applied even more strongly to many of the critics of the Preface. What is problematic in Marx's formulation is not so much the relation between the different structures, as the absence of antagonism in the base-superstructure metaphor. The only conflict mentioned in the passage is the conflict between the material productive forces of society and the existing relations of production – a conflict which, to judge from this particular passage, runs its course quite independently of human will. To modify Marx's formulation by speaking of the 'relative autonomy' of the superstructure does little to change this: the same lifeless model is simply reproduced in another shape.

The 1859 Preface can be contrasted with another passage from Marx which also emphasises the centrality of production, but does so in a very different way:

'The specific economic form, in which unpaid surplus-labour is pumped out of direct producers, determines the relationship of rulers and ruled, as it grows out of production itself and, in turn, reacts upon it as a determining element. Upon this, however, is founded the entire formation of the economic community which grows up out of the production relations themselves, thereby simultaneously its specific political form'.

(1967/71, III, p. 791)

The key here is production, no less than in the passage from the 1859 Preface, but here production is presented not as economic base, but as unceasing antagonsim. Any class society has at its core an antagonistic relation, a relation of conflict: the pumping of surplus labour out of the direct producers. The conflict never stops; if the ruling class stops pumping, the society will collapse. The form taken by this constant antagonism is the key to understanding any class society.

This passage from *Capital* gives us a starting point that is very different from that provided by the usual interpretations of the 1859

Preface. The 1859 Preface leaves us helpless, mere objects of historical change as productive forces and productive relations clash high above our heads. The passage above places us at the centre of the analysis, as part of a ceaseless class antagonism from which there is no escape, because we all relate in some way to the reproduction of society and to the pumping of surplus labour on which it depends.

Form and Fetishism

Class struggle, then, is no less central for the Marx of *Capital* than it was for the Marx who had written the *Communist Manifesto* nearly twenty years earlier. What can be seen is not a shift from class struggle to the 'laws of capitalist development', but a shift of attention from class struggle in general to the specific *form* taken by class struggle in capitalist society. The importance of *Capital* lies not in the fact that it is a study of the economic base or of the 'objective laws of capitalist development', but in the fact that it is an analysis of struggle.

This is not to say that the primary concern of *Capital* was to assert the centrality of struggle. That had already been done in the earlier works and was in any case obvious to the people for whom Marx was writing. Marx's concern is rather to understand what is specifically different about class antagonism in capitalist society. *Capital* is a formal analysis of struggle in capitalist society, an analysis of the forms taken by the antagonistic social relations. That is why, on the one hand, the clenched fist is not always obvious to the reader; but it is also why all the categories of *Capital* are categories of struggle.

The categories of *Capital* are categories of antagonism from the very beginning. This does not mean that Marx starts directly from the relation of exploitation, as Negri (1984), for example, suggests that he ought to have done: the analysis of the production of surplus-value, the form in which surplus labour is pumped from the direct producers under capitalism, does not start until chapter 7. *Capital* begins rather with the analysis of commodity and value. This has led to many economistic interpretations which have seen in *Capital* the textbook of Marxist economics – an assumption tacitly accepted even by many who are critical of economistic interpretations of Marxism (Negri 1984). Marx's argument, however, is that these categories are important not as the basis of a Marxist economics, but because they are fundamental forms in which the antagonistic social relations present themselves.

Capital begins by telling us that in capitalist society wealth presents itself to us as 'an immense accumulation of commodities', and 'a commodity is, in the first place, an object outside us'. In the apparently innocent remark that a commodity is 'an object outside us', we are presented right at the outset with the most violent antagonism of all: capitalism is the denial of our identity, the rule of things.

The commodity is not, of course, simply 'an object outside us'. In the course of the first chapter, Marx establishes that commodities are the product of human labour and that the magnitude of their value (which is the basis of the ratio in which commodities exchange) is determined by the amount of socially necessary labour required to produce them. The commodity is not an 'object outside us', it is the fruit of our (collective) work, which is the only source of its value.

Under capitalism, however, the commodity presents itself to us as, or is 'in the first place', 'an object outside us'. We neither control the things we produce, nor do we recognise them as our product. In a society in which things are produced for exchange rather than for use, the relations between the producers are established through the value of the commodities produced. Not only that, but the relation between commodities comes to take the place of the relation between the producers who produced them: the relations between producers take the form of relations between things. This Marx refers to as commodity fetishism: like gods, commodities are our own creation but appear to us as an alien force which rules our lives. Under capitalism, our lives are dominated by commodities (including money), as the form assumed by the relations between producers. The free flow of relations between people, the 'sheer unrest of life', as Hegel (1977, §46) puts it, is held captive in the fixed form of things, things which dominate us, things which shatter the unity of life into so many discrete parts and make the interconnections incomprehensible.

The labour theory of value is a theory of fetishism. In discussing commodities, Marx establishes that the value of a commodity is determined in its magnitude by the amount of socially necessary labour required to produce it. However, there is an even more fundamental question. The point is not only to understand what lies behind value, but also to understand why labour in a capitalist society takes the strange, mystified form of value. For Marx, this is what distinguishes his method from that of the classical political economists such as Smith and Ricardo. They are concerned simply to understand what determines the magnitude of value; the second question, the question of why labour takes the form of value, cannot even enter their minds since their perspectives are limited to the capitalist society in which they live. For Marx, who is looking at capitalist society as a transient society leading to a communist society in which work would be organised in a completely different way, the question of the forms assumed by the relations between producers is fundamental. The question of form is fundamental for Marx for precisely the same reason that it is invisible, a non-question, for any theory that assumes the permanence of bourgeois social relations: namely, that these forms (commodity, value, etc.) 'bear it stamped upon them in unmistakeable letters that they belong to a state of society, in which the process of production has the mastery over man, instead of being controlled by him' (1967/71, I, 81). The fact that labour is represented by value, and that social relations between producers take the form of value relations between commodities, is in itself unfreedom, the inability of people to control their own lives.

The theory of value is a theory of fetishism and the theory of fetishism is a theory of domination. The theme of *Capital* from the beginning is unfreedom: we live in a world surrounded by commodities, by 'objects outside us', which we have produced, but which we do not recognise and do not control. The forms assumed by relations between people are in themselves expressions of the fact that 'the process of production has the mastery over man, instead of being controlled by him'. The very fact that the sheer unrest of life is frozen into forms which stand opposed to people, which appear as 'objects outside us', is in itself the negation of freedom in the sense of collective self-determination.

The three volumes of *Capital* are a development on the theme of commodity fetishism. Starting from the exchange relation, Marx shows how the equality of the exchange relation conceals the exploitation of the production process, and then how layer after layer of mystification is constructed, concealing the relation of exploitation further and further from view. Capitalism is an 'enchanted, perverted, topsy-turvy world' (Marx 1967/71, III, p. 830) of fetishised forms. It is a fragmented world, in which the interconnections between people are hidden from sight. When we look at the world, it is not just through a glass darkly, but through a glass shattered into a million different fragments.

However, it is not just our perception of reality that is fragmented: reality itself is fragmented. The forms in which social relations appear under capitalism are not mere forms of appearance. It is not just that social relations appear in the fragmented form of things: social relations are in fact fragmented and mediated through things, that is, the form in which they exist. When, for example, we buy a car, the nature of the relation between the producers of the car and our own work takes the form of a relation between our money and the car: the social relation appears as a relation between things. Even after we have understood that, however, the relation between ourselves and the car workers continues to be mediated through the commodity exchange. The fragmentation of society is not only in our mind; it is established and constantly reproduced through the practices of society.

Fetishism and Class Decomposition

Marx's theory of commodity fetishism is not distinct from his theory of class. The dominant role of the commodity as the mediator of social relationships is not separate from the nature of exploitation; on the contrary, it is the fact that exploitation in capitalist society is established through the sale and purchase of labour-power as a commodity which establishes the generality of commodity relations. The core of the matter is the form 'in which unpaid surplus-labour is pumped out of direct producers'.

Like other class societies, capitalism is based on the pumping of surplus labour out of the direct producers. What distinguishes capitalist exploitation from other forms of exploitation is that it is mediated through exchange. The workers are free in the double sense of being liberated from personal ties of bondage and of having no control over the means of production: the first aspect of their freedom permits them, and the second aspect forces them, to sell their labour-power in order to survive. In return they receive the value of their labour-power in the form of a wage. The capitalist sets the workers to work and they produce a value greater than the value of their labour-power: this additional, or surplus, value is appropriated by the capitalist in the form of profit.

The fact that exploitation in a capitalist society is mediated through the sale and purchase of labour-power as a commodity conceals the class nature of the relation between capitalist and worker in at least two senses. Firstly, the relation between capital and labour is fragmented. It takes the form of so many different labour contracts between so many workers and so many employers. This not only creates divisions within capitalist units; even more, it implies divisions between the workers employed by different capitalists. The general fragmentation of social reality is reflected in the (apparent and real) fragmentation of class relations. Society appears not in the form of antagonistic classes, but as so many different groups, each with their own distinct interest. Society appears, and is, atomised and fragmented.

Secondly, the relation between capital and labour does not appear as a relation of exploitation at all, but as a relation of inequality, of (possible) unfairness. The relation of exploitation appears as a relation of exchange between (rich) employer and (poor) employee. What appears is not the direct antagonism of exploitation, the ceaseless conflict involved in pumping surplus labour out of the direct producers, but a society in which there is inequality, injustice, wealth and poverty. The relation of exploitation appears as a problem of maldistribution. Capitalist society appears to be made up of (richer or poorer) individuals rather than the unceasing antagonism between exploiting and exploited class. Struggles for social change do not take the form of an attack on exploitation, but rather calls for greater social justice, campaigns against poverty, appeals for more 'Freedom, Equality, Property and Bentham' (Marx 1967/71 I, p. 176).

Fetishism and Fetishisation

The resulting picture is a depressing one. Society is based on exploitation, on the pumping of unpaid surplus labour out of the mass of the population, yet the form taken by this exploitation has as a consequence both the fragmentation of society and the appearance of society as non-exploitative. Capitalist society presents itself to us as so many fragments, abstractly generalised in the concepts of value, money, rent, profit, state, technology, interest groups, etc. The only way of understanding the interconnection between these concepts is to see them as historically specific forms of social relations, but, as we have seen, this path is barred to bourgeois theory; not necessarily because of any dishonesty or stupidity, but simply because the concept of form makes sense only if one looks at capitalist society from the point of view of its overcoming. Inevitably, then, bourgeois theory (i.e. theory which takes for granted the continuing existence of bourgeois social relations) can only build upon the discrete forms in which social relations present themselves. Division, divide and rule, fragmentation is the principle of theoretical abstraction in bourgeois theory, as it constructs its distinct disciplines of political science, economics, sociology, law, computer science, etc. in order to understand society. The result is not to show the interconnections between the fragmented forms of social reality, but to consolidate the fragmentation. The more sociology develops its theories of groups, the more political science develops its theories of the state, the more economics develops its theory of money, so the more coherent the fragmentation of society becomes, the less penetrable its interconnections.

The interconnections are not, however, totally impenetrable. *Capital*, as a critique of bourgeois theory, is a critique of the fragmented appearance of society. The concept of form implies that there is some underlying interconnection between the forms. That interconnection is production and the way in which people relate to it, the relations of production. Underlying the self-presentation of society as a society composed of more or less equal individuals is the interconnection of these 'individuals' through production: it is the way in which production is organised that gives rise to the constitution of the individual and to the apparently chance inequalities between individuals. Underlying the fragmentation of so many different processes of production is the movement of value, the thread that binds the world together, that makes apparently quite separate processes of production mutually interdependent, that creates a link between the struggles of coal miners in Britain and the working conditions of car workers in Mexico, and *vice versa*.

However, understanding the interconnection between the fragments of society does not mean that the fragmentation is overcome; it does not 'dissipate the mist through which the social character of labour appears to us to be an objective character of the products themselves' (Marx 1967/71, I, 74), since that mist is a product of capitalist social relations. But as long as that mist exists, as long as society is fragmented, what possibility is there of radical social change? The possibility of anti-capitalist revolution presupposes that class relations must appear as such, that the fragmentation (or decomposition) of the working class must be overcome. Capitalism is a class society which does not appear to be a class society; but if it does not appear to be a class society, how can one envisage a working-class revolution? If class de-composition, as an aspect of commodity fetishism, is built into the nature of capitalist relations, how can one imagine the working-class re-composition necessary to overthrow capitalist social relations?

There are various possible responses to this dilemma, all found within the Marxist tradition. One response is that of the tragic intellectul: although we, as Marxist intellectuals, can penetrate the fetishised appearances and appreciate what is happening, the society around us is more and more fetishised. The working class has become de-composed or atomised to the point where it is no longer possible to regard it in any way as a revolutionary subject. We can, and must, protest against the exploitative, destructive society which surrounds us, but it would be quite unrealistic to be optimistic. This position, that of the Marxist intellectual as professional Cassandra, warning in vain of the disasters to come, has a long tradition dating back to the Frankfurt School and is understandably widespread at the moment.

A second response to the dilemma is to say that we, as Marxist intellectuals who have penetrated fetishised appearances, have a particular responsibility to lead the working class through the mist, to point out the interconnections, to show what lies beneath the surface. That, crudely, is the conception that underlies Lenin's distinction between revolutionary consciousness and trade union consciousness, and the consequent role that he ascribes to the revolutionary party.

What both of these responses have in common, for all their obvious differences, is the attribution of a privileged role to the intellectual. It is assumed in both cases that the fetishised impenetrability of capitalist relationships is an established fact and that it is only through intellectual activity, through reason, that we can see through the mist. The role of Marxist theory is to act as a torch, to light the way forward (or show us that there is no way forward).

However, it may be argued that the 'mist' of fetishism is not such an impenetrable fog as these theories suggest. *Capital* was a critique of bourgeois theory which showed it to be rooted in the relations of production. This does not mean that everybody is totally imbued with the conceptions of bourgeois theory. As Marx points out, the interconnections between social phenomena are clearer to 'the popular mind' than to the theorists of the bourgeoisie:

'It should not astonish us. . .that vulgar economy feels particularly at home in the estranged outward appearances of economic relations in which these prima facie absurd and perfect contradictions appear and that these relations seem the more self-evident the more their internal relationships are concealed from it, *although they are understandable to the popular mind'.*

(1967/71, III, p. 817: my emphasis)

This would suggest that the fetishised forms in which capitalist relations appear are not a totally opaque cover completely concealing class exploitation from those who are subjected to it. The apparent neutrality and fragmentation of the forms, the mystifying disconnections, come into constant conflict with the workers' experience of class oppression. Money, capital, interest, rent, profit, state – all are commonly experienced as aspects of a general system of oppression, even though their precise interconnections may not be understood. If Marx's mist metaphor is taken, perhaps it is better to see the mist not as a static, impenetrable fog, but as constantly shining mist patches. Interconnections appear and disappear, at moments the mist disperses, at moments it redescends. Fetishism is not static, but a constant process of defetishisation/refetishisation.

Seeing fetishism as a process of defetishisation/refetishisation has important consequences, theoretically and politically. The understanding of fetishism as established fact, as blanket fog, leads to a concept of revolution as event, an exogenous event which is either virtually impossible (the pessimistic position) or will be the triumphal conclusion of the growth of the party. Before the Event, capitalism is a closed system and will follow the 'laws of motion' analysed in *Capital*.

To see fetishism as a process of defetishisation/refetishisation is to emphasise the inherent fragility of capitalist social relations. Defetishisation/refetishisation is a constant struggle. The process of defetishisation, the putting together of the fragments, is simultaneously a process of class recomposition, the overcoming of the fragmentation of the working class. It is through the practical organisation and struggle of the working class that social interconnections are established, in practice and in perception. The process of refetishisation is also a process of class decomposition, a breaking down of the connections established, in practice and in theory. The survival of capital depends on the successful refetishisation/decomposition of class. The reproduction of capital is not automatic: it is achieved through struggle.

If fetishism is to be understood as a process, then this must affect the understanding of Marx's categories. The forms of social relations analysed by Marx are not closed. Value, it was seen, is not just a form of social relations, but of antagonistic social relations. But if antagonism is to have any meaning, there must be an element of uncertainty, of openness, at the core of the category. To say that social relations are antagonistic is to say that they develop through struggle, that therefore they can never be regarded as predetermined. To understand value, then, we must prise open the category, understand value as struggle, a struggle in which we are inescapably involved. To say that commodities exchange at their value is true as a generalisation, but it is certainly not a smooth and automatic process. This is not just because of the modifications which Marx introduces (the distinctions between price, price of production, value, etc.), but because in fact very often commodities are stolen. Value rests on the respect for property, and as anyone who has taken a small child to a shop, or has recently copied music or software or a book, will know, respect for property is in fact very fragile in our society. The more or less smooth operation of value is maintained in practice by an immense apparatus of education and coercion. When we say that value is a form of social relations, we must recognise the antgonism which that statement contains, the strength of the antithesis of value not only in a post-revolutionary society, but within capitalist society.

Possibly an analogy can be made between the forms of social relations analysed by Marx in the categories of value, money, rent, etc. and marriage in a traditional society. It would be true to say of such a society that marriage was the form in which sexual relations were organised in that society. However, even in the most traditional of societies, the sheer unrest of sex bursts the bounds of marriage time and time again, in thought and in practice. This is not to say that each act of extra-marital sex would be revolutionary any more than each act of theft in a capitalist society: on the contrary, it might even be seen as reinforcing the institution of marriage. But it would be clearly wrong to accept at face value the statement that marriage was the form assumed by sexual relations in that society without seeing the strength of its antithesis.

The forms of social relations analysed in *Capital* are forms which contain their own antithesis. Capitalism is a fetishised, alienated society, but the reason we can recognise it as such, and the reason we can conceive of a non-alienated, non-fetishised society, is because the antithesis of that society is contained within it. The sheer unrest of life is held captive in fetishised forms, in a series of things, but it is always there, always bursting its bounds, always forcing the fetishised forms to reconstitute themselves to keep it captive.

Our experience of capitalist society, therefore, is a very contradictory one. Coexisting with the experience of social relations as they present themselves is the opposite experience. Money is money is money, a thing. But money is also widely experienced as power, as a class relation, however vaguely. Concern about the 'fairness' of the wage contract (a fair day's pay for a fair day's work as the classic expression of fetishised trade union consciousness) coexists with very direct expressions of revolt against exploitation in the workplace. The more intense the social antagonisms, the less securely established will be the fetishised self-presentation of social relations. It is not theoretical reflection but anger born from the experience of oppression that provides the cutting edge to pierce the mystifications of capitalist society. The role of theory is not to lead the way but to follow, to focus on the contradictory nature of experience, to give more coherence to the vaguely perceived interconnections, to broadcast the lessons of struggle.

A mist that comes and goes, a constant process of defetishisation/refetishisation, might appear to be a completely unstructured conception, without direction; but that is not the case. The shifting mists do not shift without direction. The process of defetishisation/refetishisation and class recomposition/decomposition is a historical process, with certain rhythms. In crisis, the apparently smooth self-reproduction of society is interrupted. The antagonisms of society intensify; new organisation, new struggle erupts; connections long invisible reappear. Crisis expresses the defetishisation of capitalist society, the recomposition of the working class.

Crisis, Marxist Economics and Marxist Political Science

Crisis is not economic, but it presents itself as such. Crisis expresses the structural instability of capitalist social relations, the instability of the basic relation between capital and labour on which the society is based. It appears as a crisis of the economy, which may have effects on other spheres of social life.

The concept of the economy as a distinct aspect of society appears only with the advent of capitalism. In pre-capitalist times, the term referred to the affairs of the household (from the Greek *oikos* – a house), and there was no clear distinction made between household management and the economy, or between politics and economics, or between economic theory and moral philosophy. This failure to distinguish the economic as a distinct conceptual realm had its basis in the nature of pre-capitalist social relations. The relations between slave and master or between serf and lord were indistinguishably economic-and-political relations: the lord not only extracted surplus labour from the serfs but exerted judicial and 'political' authority over them as well. It is only with the advent of capitalism and the separation of exploitation (by the capitalist) from the maintenance of social order (by the state) that 'economics' (first as 'political economy') and 'politics' emerge as distinct concepts. Their consolidation as concepts (and later as university disciplines) is built upon that separation.

The concepts of 'political' and 'economic' are thus specific to capitalism. They are fetishised or superficial categories insofar as they reflect the superficial fragmentation of society. The separation of exploitation from the maintenance of order is one aspect of the 'specific . . . form in which unpaid surplus-labour is pumped out of direct producers': the fact that exploitation is mediated through the sale and purchase of labour-power as a commodity implies the separation of the immediate process of exploitation from the social coercion that is inevitably required to maintain stability in a class society. The separation of the economic and the political is thus one moment of capitalist class relations, or, in other words, the economic and the political are, by virtue of the separation which constitutes them, moments of the relation between capital and labour, specific forms of the capital relation. To take the distinction between economics and politics for granted is thus to be blind to the question of *form*, to consolidate the fetishism inherent in the concepts. Marx wrote *Capital* not as an elaboration of economic theory, nor as the basis for an alternative working-class economics, but as a *critique* of political economy which shows that the conceptions of political economy give expression to the forms of appearance of capitalist class relations.

It is self-contradictory, then, to speak of a Marxist economics or a Marxist political economy. Marx's critique was a critique not just of specific theories, but of the construction of a theory on the basis of the superficial forms in which class relations appear. Economics is the study of things, of the forces (like money, value, rent, interest, etc.) which rule people's lives, and it treats those things as such, rather than as forms of social relations which 'bear it stamped upon them in unmistakeable letters that they belong to a state of society in which the process of production has the mastery over man, instead of being controlled by him'. By treating its categories as things, rather than as expressions of social relations, economics inevitably treats people as abstract, passive objects of social change.

Bourgeois economics accepts its categories as they present themselves. Money is money is money; the task for economics is to understand its relation with other economic categories, the laws governing the movement of money, etc. Because money is not seen as a form of social relations, no attempt is made to prise open the categories, to reveal 'the origins of economic developments in the concrete activities of men and women engaged in social life' (Clarke 1980, p. 5).

Yet fetishism does not disappear with the critique of its existence, nor does the conceptual hold of bourgeois thought disappear with the recognition that it is superficial. Marx's own use of the term 'economics' is ambiguous or contradictory (as in the 1859 Preface, for example), and the tradition of 'Marxist economics' is a strong one, smoothing, as it does, the contradictions between reading *Capital* and working within a university environment.

If we look at value and crisis through the eyes of Marxist economics, we get a different picture from that so far presented. Many of the assumptions of bourgeois theory are carried over into the discussion of Marxist categories once these categories are seen as economic. The categories remain closed. Although it is noted that value is a social relation, that it is specific to capitalism and will have no place in a socialist society, it is still assumed that, within the confines of capitalism, value can be treated as an economic category. Thus, for example, in discussing value, much more attention is paid typically to the magnitude of value, and the question of form is relatively neglected. This is true not only of the socalled neo-Ricardians, but also of theorists who would describe themselves and be widely accepted as being Marxist. Typically, the law of value is seen as showing 'how the varying amounts of socially necessary labour required to produce commodities regulate prices' (Itoh 1980, p. 132). The critique of value as form is lost, the rigidities of bourgeois thought retained. Although value is said to be a social relation, the social aspect of it is kept in the background to be released after the revolution, 'once the direct producers are restored as the subject, rather than the object of production' (Itoh 1980, pp. 135–6). If the workers are nothing more than the object of production, if, by implication, fetishism is total, then Itoh is quite right. There is no need to prise open the category of value (apart from seeing it in historical perspective), and capitalism can be understood in term of its 'laws of motion'. But, if the workers are nothing more than the objects of production, revolution would seem to be conceptually and theoretically impossible, or rather, the only way of thinking about revolution is as an external event.

These assumptions are reflected in much of the discussion of the Marxist theory of crisis. What distinguishes Marxism from other forms of radical thought, it has been suggested above, is not so much its analysis of capitalist oppression or its vision of socialism, as the fact that it is a theory of capitalist instability. Capitalism is oppressive, but it is a self-contradictory and unstable form of oppression. A theory of crisis is a theory of this instability, and therefore a theory of the volatility of class relations. Many of the discussions of crisis, however, treat it as external to the question of class relations and class struggle. At best, the analysis of crisis provides a framework within which struggle takes place, a reminder of the mortality of capitalism, but not a theory of class relations. It is argued, for example, that crisis is made inevitable by the operation of the law of the tendency of the rate of profit to fall; that crisis involves an intensification of class struggle and may provide opportunities for revolution; but crisis as such and the tendency of the rate of profit to fall are still analysed as being economic processes, separate from class struggle. As O'Connor comments, 'the emphasis, at least, of traditional theory is that human labour power is successfully treated as if it were merely an object of exchange and labour, and that workers thus have little or no power to reverse, much less redefine, the process of self-expanding capital except in the event of a socialist revolution' (1987, p. 91).

Paradoxically, then, what should be a theory of capitalist instability becomes a theory of capitalist reproduction. Often this acquires very functionalist overtones: capitalist reproduction becomes a closed circle – until the exogenous moment of socialist revolution, of course. The laws of motion of capitalism prescribe a certain course of evolution, and until the day of revolution, workers are the objects of domination, no more.

In recent years there has been an attempt to break away from the determinism and functionalism of the Marxist economics tradition by seeking to develop a 'Marxist political theory'.

The attempt to develop a distinct Marxist political theory is rooted in the critique of Marx's statement of his method in the 1859 Preface, discussed above. As long as the 1859 Preface was taken to be the definitive statement of the Marxist method, as it was for many years by the Marxist 'orthodoxy' of the Communist Parties, then the theoretical discussion of the state was relatively neglected, since the political was regarded as simply part of the superstructure. With the crisis of the Communist Party orthodoxy from the 1960s onwards, however, the 1859 Preface was criticised for not allowing sufficient autonomy to the superstructure, and particularly to the political and ideological levels. Poulantzas, in particular, argued influentially that the relative autonomy of the different levels allowed one to develop a distinct Marxist political science, to complement the Marxist economics developed by Marx in *Capital*. In this view, the problem with the tradition of Marxist economics is that it is incomplete, and given too much weight. The logic of this approach is to say that Marxism should develop from being an economic theory of society to being an interdisciplinary theory of society (the economic still being determinant in the last instance, of course).

The problem with an interdisciplinary approach to Marxism is that it simply adds fetishism to fetishism. The Marxist economics approach is not incomplete, but superficial, in the sense that it takes for granted the separation of social relations into economic and political relations. To complement that approach with an analysis of the political which similarly takes 'the political' as an assumed starting point for the analysis is to multiply the superficiality, to further hide from view the social relations which are so fragmented. To say, for example, that a crisis is not just economic, but economic and political, is unhelpful unless the nature of the economic and of the political are also questioned. In practice, what often happens is that the 'political' analyses simply accept as given the framework provided by the analysis of the economists.

The functionalist assumptions of much of Marxist analysis (and particularly of the Marxist economics tradition) are often carried over into the notion of crisis itself. Crisis, as was seen earlier, means not just a breakdown but a turning point, an intensification of change. There are two sides to the theory of crisis, both of them present in Marx's discussion in *Capital*. On the one hand, crisis expresses the breakdown of a pattern of accumulation and confronts capital with intimations of its own mortality: the falling rate of profit 'testifies to the limitations and to the merely historical, transitory character of the capitalist mode of production' (1969/71, III, p. 242). On the other hand, crisis forces through a restructuring of capital: through the destruction of less efficient capitals and through the increase in exploitation the basis is laid for a new period of capital accumulation. Crisis is both breakdown and restructuring, both instability and restabilisation of class relations. The problem is how we understand the relation between these two faces.

Different sides of the crisis tend to be emphasised at different times. In the late 1960s and the early 1970s, when it still was not clear for all to see that Keynes had not resolved the problem of capitalist crisis, the emphasis in discussions of crisis tended to be on its inevitability, with crisis conceptualised as a rupture in the process of accumulation. As the crisis became manifest and it was clear that revolution was not imminent, the emphasis in the discussion shifted to seeing crisis as a process of restructuring and trying to understand current changes in society in terms of the restructuring of capital. What got lost in the shift from one emphasis to another was the question of the relation between the two faces of crisis, breakdown and restructuring.

Frequently, it is assumed that these two faces are in fact identical and inseparable. The destruction of one pattern of accumulation is the creation of the basis for another: the crisis is a 'creative destruction', to borrow a phrase from Schumpeter (Perez 1983, p. 159). Schumpeter is relevant here because, in Negri's view (although Negri is far from being an orthodox Marxist economist he accepts many of the assumptions of Marxist economics), it was Schumpeter who realised for the bourgeoisie what Marx had already seen many years before, that crisis is an integral part of capitalist development (Negri 1968/1988). There appears to be an assumption, both in Negri's argument and in the views of many other writers on crisis theory, that crisis is a process of 'creative destruction', that the two aspects of the crisis can simply be elided. But this is precisely to extend the functionalism of bourgeois economics: if crisis is inevitably also a restructuring of capital, then the reproduction of capital is indeed a closed circle from which there is no escape.

Crisis, Fetishism and Class Composition

The two aspects of the crisis are not identical: between crisis-as-rupture and crisis-as-restructuring there is a whole world-history of struggle.

Crisis is first of all rupture, a break in the established pattern of class relations. Before the crisis it appears for a while that the world has gained stability, that major problems have been solved, that class struggle is a thing of the past. Certain things become accepted as 'normal': patterns of international relations, patterns of political conflict, patterns of occupational structure and working-class organisation, patterns of relations between women and men, between adults and children, patterns of cultural expression. Contained conflict appears as harmony. And then there is a break: conflict becomes manifest, the 'normal' is questioned, other views of this normality gain force, hidden interconnections appear, established patterns of power are attacked. The dam bursts. Suppressed anger is suppressed no longer.

Any class society, any society in which the majority of the population is subordinte in their everyday activity to the interests of the minority, is unstable. The whole of history is punctuated with revolt, since long before capitalism or capitalist crisis came into the world. Under capitalism, however, the rupture of established patterns follows a certain rough rhythm, reflected in theories of crisis, business cycles, long waves, etc. The accumulation of suppressed anger finds vent in these upheavals, but this does not explain the rhythmical regularity of crisis. The instability that is a feature of any class society takes a peculiar form under capitalism, that can be explained only in terms of the peculiarities of the relation between capital and labour. In order to understand crisis as a break in the pattern of domination, it is not sufficient to explain it simply in terms of the relations between capitals (as disproportionality theories do), or simply in terms of the patterns of distribution in society (as underconsumptionist theories do). In order to understand crisis as an expression of the peculiar structural instability of capitalism as a form of class domination, it is necessary to seek a flaw, a geological fault, as it were, in the relation of exploitation itself, in 'the specific . . . form in which unpaid surplus-labour is pumped out of direct producers'.

Marx analysed this fundamental instability in the capital relation in his analysis of surplus value. Capitalists, unlike the ruling class in any previous class society, are constantly driven through competition to increase exploitation, to increase the amount of unpaid surplus labour that is pumped out of the direct producers. It is this 'werewolf's hunger for surplus-labour' (1967/71, I, p. 233) which gives to capitalism its peculiar instability. As Marx analyses it in *Capital*, the capitalist greed for surplus-value takes two basic forms. The first is absolute surplus-value, the struggle by capital to lengthen the working day in order to increase the surplus-value produced. This reaches a point at which the very survival of the workers is threatened, and therefore also the survival of capital. The enactment of factory legislation to limit the length of the working day forces capital to try to satisfy its greed in another way. Instead of constantly lengthening the working day it seeks to reduce that proportion of the working day that reproduces the value of the worker's own labour-power. This is achieved basically through technological innovation and the pursuit of efficiency. As commodities are produced more quickly, the magnitude of their value (which is determined by the socially necessary labour-time required to produce them) falls also. To the extent that the commodities consumed by the workers fall in value, the labour-power itself falls in value, even though living standards may be rising. As a result, with a working day of stable length, less time is spent producing the value equivalent of the worker's labour-power, and more time spent on producing surplus-value. This form of maximising surplus-value is referred to by Marx as relative surplus-value.

Relative surplus-value implies the constant pursuit of technological innovation and the constant reorganisation of the production process. It implies also a change in the relation between living labour (the worker in action) and dead labour (the machinery and raw materials, the product of past labour): as technology progresses, there is a tendency for each worker to move a greater and greater mass of machinery and raw materials. In terms of the composition of capital, this tends to express itself as a relative rise in that part of the capital invested in constant capital (machinery and raw materials) and a relative fall in variable capital (the part of the capital invested in the purchase of labour-power): there is, as Marx puts it, a rising organic composition of capital.

The pursuit of relative surplus-value, then, means that capital is never at peace. It is always restless, always seeking change, unlike the ruling classes in previous class societies. It is also constantly expelling from the production process, in relative terms, the only source of its own existence, living labour. It is only living labour that produces value and, as capital becomes more encumbered with dead labour, the proportion between the surplus-value produced (by the living labour) and the total investment of the capitalist tends to fall.

In other words, the pursuit of relative surplus value is associated with a tendency for the rate of profit to fall.

The tendency for the rate of profit to fall, analysed by Marx in the third volume of *Capital*, is, therefore, an economic manifestation of constant changes in the organisation of the production process. The same changes ensure that the antagonism between labour and capital is kept very much alive. Resistance and latent revolt are inherent in any relation of subordination. Even between the meekest slave and the most dominant master there is an active antagonism, a (perhaps unexpressed) tension of mutual dependence, that makes the relation dynamic. The dependence of capital on constant changes in production, on constantly pursuing increases in surplus-value, ensures that the antagonism between labour and capital is kept open and constant, even in periods of relative stability. The workers organise, defensively and offensively; the capitalists' struggle to remain in control is inseparable from the struggle to maximise surplus value. Relative surplus value expresses itself in the dynamic of class struggle, in the changing forms of attack and counter-attack, in the changing composition of both labour and capital. Here again, the peculiarly unstable dynamic of capitalism comes to the fore. The more capital is successful in accumulating surplus value, the more labour grows as a destructive force in its midst. A period of successful accumulation is expressed potentially in the growth of working-class strength and organisation, as unemployment is reduced and the bargaining position of labour is strengthened. The more successful capital is, the more the one fundamental contradiction of its existence comes to the fore: its dependence on labour. All masters depend for their existence on their servants. In the case of capital, this fundamental fact of life is forced upon it just when it feels itself strongest.

Relative surplus-value production carries within it its own destructive force, manifested both in the tendency for the rate of profit to fall and in the growing class composition of the working class. As a period of rapid accumulation progresses, there is a tendency for the working class to grow in organisational strength and combativity, and for the rate of profit to fall. It becomes more difficult for capitalists either to get the rate of profit they expect or to reorganise the process of production in the way that they wish. Antagonisms intensify, the contradictions of capitalism become more obvious, the interconnections between previously discrete phenomena become plainer, capitalism as a form of social organisation is criticised more widely and more openly.

Capitalism is seen to be in crisis. This is perceived as a crisis of the economy: profits fall, competition intensifies, firms go bankrupt, whole sectors and whole countries decline. But it is not just seen as economic. It is seen as a crisis of the state: where, before, the state appeared to be able to ensure the smooth development of society, it now seems incapable of doing so. But it is also a crisis of the family, of morality, of religion, of trade union structures, of everything that previously seemed to ensure social harmony and is now no longer able to do so. There is a strong feeling in the capitalist class that things cannot continue as they are. The constant process of change inherent in capitalism is now seen as being insufficient: something more radical is required. The previous process of change is seen as being part of a pattern, and it becomes clear that that pattern has come to its end.

This is crisis: a breakdown in the established pattern of social relations. To the capitalist class, the future seems uncertain, dangerous. There is no obvious way forward, apart from attacking the strength of organised labour and anybody deemed subversive, and calling for a return to morality, discipline, order. This is not restructuring: it is rupture.

Clearly, the rupture may contain the possibility of a restructuring. For parts of the capitalist class, there is no future: bankruptcies increase sharply, political parties associated with the previous social pattern go into irreversible decline. However, new industries may take the place of the old, new political parties may arise, the economic decline of one country may be balanced by the rise of another. The way forward is not clear, but there are all sorts of experiments in new forms of management, new forms of technology, new relations between state and industry, new patterns of political organisation. It may be that the composition of the working class can effectively be broken by a combination of violence, legal restriction and economic reorganisation. It may be that capital will then be able to impose all the changes in production that it desires. All that may be, but it is not pre-determined.

It may be that after some time, capital becomes more confident about the future, that it is possible to discern the possible bases of a new relatively stable pattern of accumulation. That is the situation in which we are at the moment. It is at this point that 'Marxist' functionalism becomes most insidious. There is a whole new world of academic analysis opened up: to theorise the new patterns of accumulation, to give a name to the new form of domination, and in so doing to consolidate it. Crisis-as-rupture is forgotten, or remembered only in so far as rupture is seen as a preliminary phase of restructuring. The new patterns are seen as established, having 'emerged' as a new reality which has to be accommodated, rather than as a project, which capital has yet to impose through hard struggle. From being a theory of struggle, Marxism, once struggle is forgotten, easily becomes a theory of domination.

That cannot be. The breakdown of a pattern of social relations does not imply either its immediate or its successful restructuring. It may be that rupture contains the possibility of restructuring. It may be that that possibility is realised, as it has been in the past. But that is not certain, even now, and if a new pattern of relatively stable capitalist social relations is established, it will not simply emerge but be the result of a long and very bloody struggle. Between crisis-as-rupture and crisis-as-restructuring there is an abyss of possibility, a *salto mortale* for capital with no guarantee of a safe landing, a whole history of the world in struggle.

References

Clarke, S. (1980), 'The Value of Value: Rereading *Capital*', *Capital & Class*, no. 10.

Hegel, G. W. F. (1977) *The Phenomenology of Spirit* (Oxford)

Itoh, M. (1980) *Value and Crisis* (London).

Marx, K. (1967/71) *Capital* (Moscow).

Marx, K. (1971) *Contribution to the Critique of Political Economy* (London).

Negri, A. (1984) *Marx Beyond Marx* (South Hadley, Mass.).

Negri, A. (1968/1988) 'Marx on Cycle and Crisis', in Negri *Revolution Retrieved* (London).

O'Connor, J. (1987) *The Meaning of Crisis* (Oxford, New York).

Perez, C. (1983) 'Structural Change and Assimilation of New Technologies in the Economic and Social Systems', *Futures* (October).

Rader, M. (1979) *Marx's Interpretation of History* (New York).

Stern, R. (1970) 'Historians and Crisis', *Past and Present* no. 52.

Index

References to Marx are omitted since he is discussed throughout the text.

Panzieri, R., xvii, 36, 111, 112, 128, 136
Parmenides, 26
Pascal, B., 38
Pascal, R., 11
Perez, C., 164
Picciotto, S., *see*: Holloway, J. and S. Picciotto
Plamenatz, J., 35, 36
Plato, 3, 6, 15, 18, 26
Polanyi, M., 54, 66, 119, 132, 133, 139
Popper, K. R., xviii, 15, 27, 49, 50
Poulantzas, N., 36, 163
Prinz, A., 37
Proudhon, P.-J., 19
Psychopedis, K., xv, 67

Rachleff, P., 141
Rader, M., 145
Reichelt, H., 61, 67
Reid, T., 38
Ricardo, D., 78, 152
Robespierre, M. M. I., 97
Roemer, J., xviii, 9, 34
Rorty, R., xviii
Rose, G., 1
Rouffignac, A. L. de, 137, 143
Rousseau, J.-J., 29
Russell, B., 141
Ryan, C., xi, xvi

Sartre, J.-P., 10, 12, 14, 40
Say, W. B., 47
Schmidt, A., 28, 61, 67
Schmidt, C., 35
Schumpeter, J., 164
Senior, N. W., 135
Shanin, T., 139, 141
Shipway, M., 141
Skinner, A., 11
Smith, A., 111, 118, 152
Socrates, 6, 15
Sohn-Rethel, A., 139
Spengler, O., 46
Spinoza, B., 21, 27, 35
Stalin, J., xiii, 97
Stern, R., 146
Sweezy, P., 136

Taylor, F., 118, 120
Thales, 27
Tomassini, R., 137
Tronti, M., xvii, 104, 111, 115, 128, 130, 137, 141, 142

Walsh, W. H., 7
Wartofsky, M., 48
Weber, M., 38, 103
Williams, R., 24
Wittfogel, K. A., 10
Wright, E. O., xviii